Silver Educational Publishing
Published by Silver 8 Production LLC

ISBN 1-933023-00-7

D0074330

SILVER EDUCATIONAL PUBLISHING
SEPBOOKS.COM

The Ultimate Study Guide For Biology:
Key Review Questions and Answers with Explanations

Topics
Origin of Living Things &
Chemistry of Life
Structure and Function of the Cell &
Energy Pathways
Reproduction and Heredity
Genetics

Volume 1
Dr. Patrick Leonardi

FOREWORD

This is volume 1 of a three volume series. There are 322 questions, which are quite extensive for each topic of origin of living things & chemistry of life, structure and function of the cell & energy pathways, reproduction & heredity and genetics. Make sure to also purchase the following:

THE ULTIMATE STUDY GUIDE FOR BIOLOGY: *KEY REVIEW QUESTIONS AND ANSWERS WITH EXPLANATIONS (TOPICS: EVOLUTION, ECOLOGY, KINGDOM BACTERIA, KINGDOM PROTISTA, KINGDOM FUNGI & VIRUSES, PLANT FORM AND FUNCTION) VOLUME 2*

THE ULTIMATE STUDY GUIDE FOR BIOLOGY: *KEY REVIEW QUESTIONS AND ANSWERS WITH EXPLANATIONS (TOPICS: KINGDOM ANIMALIA, ORGANIZATION OF THE ANIMAL BODY, ANIMAL FORM AND FUNCTION, ANIMAL REPRODUCTION, DEVELOPMENT AND BEHAVIOR) VOLUME 3*

Other Titles Available From
SILVER EDUCATIONAL PUBLISHING (SEPBOOKS.COM)

Anatomy and Physiology Study Guide: Key Review Questions and Answers with Explanations (Volume 1)

Anatomy and Physiology Study Guide: Key Review Questions and Answers with Explanations (Volume 2)

Key Facts for Anatomy and Physiology

Microbiology Study Guide: Key Review Questions and Answers

Histology Study Guide: Key Review Questions and Answers

Contents

Origin of Living Things & Chemistry of Life

1. Which of the following is not a characteristic of all life?
       a. sensitivity
       b. reproduction
       c. regulation
*   d. deductive reasoning

Sensitivity refers to the ability of all living things to react to a particular stimulus. All living things reproduce and have regulatory aspects that make sure their internal aspects function properly. Deductive reasoning does not apply to all living things.

2. The creation of new types of living things from the process of natural selection is called _____.
*   a. evolution
       b. reproduction
       c. biological diversity
       d. regulation

3. The atoms of a specific element have a different number of neutrons. For example, the element hydrogen might have 2 neutrons and another hydrogen might have 3 neutrons. This different number of neutrons refers to the _____ of this element.
       a. proton
       b. atomic mass
*   c. isotope
       d. half-life

A proton has a positive charge. The number of protons usually equals the number of electrons in an atom. Electrons are negatively charged. Atomic mass of an atom equals the total amount of its protons and neutrons. Half-life refers to how long it takes for 50% of an isotope to decay.

4. Bones in the leg of two different vertebrates have different structure and function. In addition, these bones have the same evolutionary beginning. This best describes _____ .
       a. biology
       b. artificial selection
       c. analogous bones

*   d. homologous bones

5. _____ has less electrons than protons; it has a _____ charge.
*   a. Cation, positive
        b. Anion, positive
        c. Anion, negative
        d. Cation, negative

An atom that has more electrons than protons is referred to as an anion. An anion has a negative charge. An atom has a nucleus that is made up of neutrons and protons. However, it is surrounded by an orbit or orbits of electrons.

6. Natural selection was an evolutionary theory written by _____ and _____.
        a. Darwin, Bohr
*   b. Darwin, Wallace
        c. Malthus, Bohr
        d. Wallace, Bohr

7. Which of the following is a polysaccharide that consists of glucose in the beta form and makes up the cell walls of plants?
        a. prostaglandins
        b. triglyceride
*   c. cellulose
        d. terpenes

The alpha form of glucose is composed of starch. Most living things cannot digest cellulose. This is because most organisms lack an enzyme required to break down molecules between two beta-glucose molecules. Prostaglandins make up a group of approximately 20 lipids that are in vertebrate tissues. A triglyceride consists of three fatty acids. Terpene is a type of lipid that can be found in the pigment of eyes in vertebrates; it is also found in the chlorophyll of plants.

8. Which of the following is not true of glycerol?
        a. is a three carbon alcohol
        b. it is the backbone of the structure of fat
        c. it has a hydroxyl group
*   d. it is a three carbon sugar

9. The slow change in populations from one generation to future generations most closely describes _____.
*   a. evolution
        b. artificial selection
        c. survival of the fittest
        d. biology

Charles Darwin was credited with his survival of the fittest theory of evolution. He explained that animals were different in different locations. This was because different places and climates required animals to have different characteristics to survive. If certain animals could not adapt to the environment, extinction would result. Only the animals that could adapt would survive. Charles Darwin also had explained artificial selection as when animal breeders would breed for characteristics most desirable. Natural selection occurs when animals with appropriate characteristics for survival to their specific environment will grow to be more widespread in the future.

10. Most of the weight of the human body is made up of which of the following elements?
        a. nitrogen
        b. carbon
        c. hydrogen
*   d. oxygen

11. Which is not a component of an amino acid?
*   a. glycerol
        b. carboxyl group
        c. central carbon atom
        d. amino group

Amino acids make up proteins. In facts, proteins can be made up of up to 20 various amino acids. A carboxyl group is chemically written as COOH. An amino group is chemically written as NH_2. Glycerol is a three-carbon alcohol and makes up a part of the fat molecule.

12. What is called a covalent bond that connects two amino acids?
        a. fatty acid
        b. glycerol
*   c. peptide bond

d. terpene

13. Which of the following is a modified fat that has two fatty acid chains coupled with no attachment to a five-carbon ring?
*   a. phospholipid
       b. glycerol
       c. triacylglycerol
       d. prostaglandins

Triacylglycerol is also known as triglyceride.   Having too many triglycerides in the body can lead to heart disease and stroke. Elevated levels of triglycerides clog up the arteries. Prostaglandins are lipids that consist of fatty acids attached to a five-carbon ring. Phospholipids in water have a lipid bilayer that consists of hydrophobic and hydrophilic portions. Hydrophobic refers to "water fearing or repelling" and hydrophilic refers to "water loving."

14. A contractile protein plays a key role in _____ and _____ is an example.
       a. support, insulin
*   b. motion, actin
       c. enzyme catalysis, keratin
       d. support, kinase

15. Which of the following is not a component of carbohydrates?
       a. hydrogen
       b. oxygen
*   c. nitrogen
       d. carbon

Carbohydrates consist of monosaccharides, disaccharides, and polysaccharides. Monosaccharides are known as simple sugars. They have an empirical formula of $C_6H_{12}O_6$. Glucose, a six-carbon sugar is an example of a monosaccharide.   Disaccharides are composed of two monosaccharides connected by a covalent bond.   Polysaccharides are the large molecules composed of monosaccharide subunits. Polysaccharides consist of many glucose molecules connected into chains.

16. Which of the following is an example of a polysaccharide?
*   a. cellulose

b. fructose

c. galactose

d. triglyceride

17. Galactose is considered a(n) _____ of glucose. In addition, fructose is a(n) _____ of glucose.

    a. isomer, isomer

    b. isomer, stereoisomer

*   c. stereoisomer, isomer

    d. stereoisomer, stereoisomer

Sugars that are isomers have the same empirical formula. Fructose and glucose have the same empirical formula of $C_6H_{12}O_6$. Fructose is really just another form of glucose. The double bonded oxygen in fructose is just attached to an inner carbon. In glucose, the double bonded oxygen is attached to the terminal carbon. Glucose and galactose are mirror images of each other, in one aspect, and that is in regards to an OH group. Galactose has this OH on one side of its molecule and glucose has the OH group on the other side of its molecule. However, remember this OH group will appear as HO on the galactose chemical structure and OH on the glucose chemical structure. This is what makes them stereoisomers.

18. A starch, such as amylose, is an example of which of the following?

    a. monosaccharide

*   b. polysaccharide

    c. prostaglandin

    d. terpene

19. Which of the following makes up the external skeleton of insects and crustaceans?

    a. glycerol

*   b. chitin

    c. triglyceride

    d. terpene

Chitin is a changed form of cellulose that is composed of glucose units with one nitrogen added. Chitin is a structural carbohydrate.

20. Enzymes are classified under which type?

*   a. proteins

b. lipids

c. fatty acids

d. disaccharides

21. The process of adding a molecule of water in a catabolic reaction in order to break a covalent bond occurs in:

    a. dehydration synthesis

    b. denaturation

*   c. hydrolysis

    d. oxidation

Dehydration synthesis is essentially a chemical reaction where water is lost. Denaturation is the process by which protein alters its form due to a changing environment. In chemical reactions, oxidation refers to the loss of an electron; reduction refers to the gaining of an electron.

22. Which of the following is a false statement?

    a. An outer energy level of an element can have less than nine electrons.

    b. It is possible for an element to have one electron in its outer energy level.

*   c. Inert elements never have eight electrons in their outer energy level.

    d. Most elements are inclined to full their outer energy levels.

23. Which of the following most closely explains the Octet rule?

*   a. Most elements are inclined to full their outer energy levels.

    b. It is possible for an element to have one electron in its outer energy level.

    c. Inert elements are quite reactive.

    d. The loss of an electron is referred to as oxidation.

Inert elements have their outer energy level filled with eight electrons. Inert elements are not reactive. Helium and radon are examples of inert elements. However, elements with just one electron in its outer energy level are highly reactive. Valence electrons are electrons in the outer energy level of an element.

24. Which of the following is created when two atoms share one or more pairs of electrons?

a. oxidation
b. reduction
c. ionic bond
* d. covalent bond

25. Which of the following is not a type of covalent bond?
* a. ionic bond
b. single bond
c. double bond
d. triple bond

Covalent bonds can either by single, double or triple bonds. Triple bonds are the strongest, followed by double bonds and finally, single bonds are the weakest. Ionic bonds occur when two opposite charged atoms are attracted to each other. In fact, the unpaired electron in the outer energy level of an atom will go to another atom. Since the original atom loses an electron, it now becomes positively charged. The atom that gains an electron will now become negatively charged.

26. Which of the following increases the rate of a chemical reaction?
a. actin
b. myosin
c. ionic bond
* d. catalyst

27. Salt water is a(n) _____ and has a pH _____ 7. It has a _____ concentration of hydrogen ions (H+).
a. base, below, low
* b. base, above, low
c. acid, above, low
d. base, above, high

Water has a pH of 7, which is considered neutral. Anything below a pH of 7 is an acid and anything above a pH of 7 is considered a base. Examples of acids would be grapefruits and lemons. Examples of bases would be baking soda and salt water. A solution with an increased amount of +H ions is basic. A solution with a decreased amount of +H ions is acidic.

28. _____ serves as a storage area for hydrogen ions in the human body. When the concentration of hydrogen ions is low in a specific area of

the body like the blood, this substance gives more hydrogen ions to this area.

     a. Covalent bond
*   b. Buffer
     c. Chitin
     d. Nitrogen

29. Archaebacteria that lived in extreme salty environments are known as which of the following?
*   a. halophiles
     b. thermophiles
     c. peptidoglycans
     d. methanogens

Thermophiles are types of bacteria that like to live in hot environments. Peptidoglycan is composed of protein and carbohydrates that is present in the cell walls of the bacteria of today. Archeabacteria are ancient forms of bacteria. Methanogens produce methane and are primitive bacteria that still exist today. Methanogens reproduce without air; they are anaerobic. Methanogens are types of archeabacteria.

30. Which of the following never lacks a peptidoglycan cell wall?
     a. methanogen
     b. thermophile
     c. halophile
*   d. cyanobacteria

31. Which of the following is false regarding water?
     a. Water is held collectively together by hydrogen bonds.
     b. There are two covalent bonds in water.
*   c. One double covalent bond holds water together.
     d. Electronegativity is present in a molecule of water.

Water that is held collectively together by hydrogen bonds is called cohesion. Water is held together by two, single covalent bonds. There is one single covalent bond between one hydrogen atom and oxygen. The other covalent bond is between the other hydrogen atom and the same oxygen. Hydrogen is positively charged and oxygen is negatively charged in the water molecule. Electronegativity is exhibited more by the oxygen atom in a water molecule. This is because the oxygen atom attracts more

electrons to itself. Polar molecules, like water, are like a magnet. The positively charged hydrogen atoms and negatively charged oxygen atoms are attracted to each other without touching each other.

32. When sugar dissolves in water, it doesn't reform back to its solid form. Instead, _____ prevent this from occurring.
*   a. hydration shells
    b. triple bonds
    c. isotopes
    d. hydrocarbons

33. Which of the following is a branched polysaccharide made from plants?
    a. amylose
    b. glycogen
*   c. amylopectin
    d. glucose
Amylose is a starch in plants that is not branched and consists of many chains of glucose. Glycogen is formed in animals and consists of polysaccharides made of amylose chains that are branched. Glucose, by itself, is considered a monosaccharide. Putting many glucose subunits together will form polysaccharides. Amylopectin is a plant starch that consists of branches of amylose.

34. Biologists using principles to forecast certain results is referred to as:
    a. inductive reasoning
*   b. deductive reasoning
    c. hypothesis
    d. variable

35. When there is a double bond between two carbon atoms in a fatty acid, it is most likely _____.
*   a. unsaturated
    b. saturated
    c. glycerol
    d. amino acid
Unsaturated fatty acids contain one or more double bonds between carbon atoms. Also, there are open spots on the carbon atoms where there is no hydrogen bonded. Remember that saturated and unsaturated fatty acids are

long hydrocarbon chains. Saturated fatty acids are long hydrocarbon chains where the carbon atoms are bonded by single bonds. Also, all the slots are filled around carbon. In other words, all the carbon molecules are bonded with hydrogen atoms. In addition, both saturated and unsaturated fatty acids have COOH groups at the end of their chains.

36. When there is more than one double bond between two carbon atoms in a fatty acid, it most accurately referred to as:
*   a. polyunsaturated
      b. saturated
      c. glycerol
      d. enzymes

37. Which of the following is not an aromatic amino acid?
      a. tryrosine
      b. phenylalanine
      c. tryptophan
*   d. alanine
There are 20 common types of amino acids. All amino acids have the following things in common: carboyxl group (COOH), amino group (NH_2), and a hydrogen atom, which are all connected to a carbon atom. However, since a carbon atom has four possible bonds it can form on each side, there is one side open. This side that is open is referred to as R. In other words, every amino acid has a different R or molecule that it has. This R group is what the difference is between aromatic amino acids such as tyrosine and tryptophan as examples. Aromatic amino acids have R groups that contain organic carbon rings.

38. All of the following are nonpolar, nonaromatic proteins except:
      a. leucine
      b. alanine
      c. valine
*   d. glycine

39. Which of the following is a type of bacteria that lacks a nucleus?
      a. eukaryote
*   b. prokaryote
      c. terpene

d. Pelomyxa palustris

Prokaryotes were the first types of bacteria that are known to appear on earth. They had little internal structure and no nucleus. Eukaryotes are types of bacteria that have a nucleus. Pelomyxa palustris is an eukaryote. Terpene is a type of lipid that makes up pigments in plants for chlorophyll; it also is a pigment found in the eyes of animals.

40. Which is not part of the structure of DNA?
    a. phosphodiester bond
    b. 5-carbon sugar
*   c. 6-carbon glucose
    d. nitrogenous base

41. Which of the following is an ionizable, nonaromatic amino acid?
    a. glycine
    b. tyrosine
*   c. arginine
    d. leucine

Glycine is a polar, uncharged amino acid and is nonaromatic. There are four other polar, uncharged amino acids that are nonaromatic, which are the following: serine, threonine, asparagine, and glutamine. Tyrosine is an aromatic amino acid. Phenylalanine and tryptophan are the other aromatic amino acids. Arginine, lysine, histidine, glutamic acid and aspartic acid are types of ionizable, nonaromatic amino acids. Leucine is a nonpolar, nonaromatic amino acid along with alanine, valine and isoleucine.

42. Proline, cysteine and methionine can be described as:
    a. aromatic amino acids
    b. nonaromatic, nonpolar amino acids
*   c. special structural amino acids
    d. ionizable, nonaromatic amino

43. Which of the following plays a significant part in the transport of sugars?
*   a. disaccharide
    b. chitin
    c. casein
    d. calmodulin

Fructose and glucose form the disaccharide sucrose. Remember that two monosaccharides make up one disaccharide. Calmodulin, a protein, has a binding effect on calcium ions. Calmodulin also plays a role in activating certain enzymes to bring on a muscle contraction in the human body. Casein, a protein, stores ions in milk. Chitin forms the outside, hard skeleton of lobsters and insects. Chitin is composed of structural carbohydrates.

44. Most synthetic sugars are composed of which type?
*   a. left-handed sugars
      b. right-handed sugars
      c. sucrose
      d. fructose

45. Immunoglobulins are types of proteins found in the human body that serve as which function?
      a. regulation
      b. support
*   c. defense
      d. enzyme catalysis

Regulation proteins are types of proteins that regulate functions in the body. For example, insulin is secreted by the pancreas in response to increased blood sugar levels. Insulin has a function of lowering blood sugar levels. Collagen is the most abundant protein in the body and serves as a function of support in animals. Collagen is found in ligaments, skin, tendons and bones. Immunoglobulins are proteins that play a role in immunity. Antibodies, which are immunoglobulins get rid of foreign substances in our bodies such as bacteria and viruses. Enzymes break down complex substances into simpler ones to be used as energy by the body. Kinase is an example of an enzyme.

46. Which best describes a peptide bond?
      a. a hydrogen bond that connects two lipids
      b. an oxygen bond that connects two hydrogen atoms
      c. an ionic bond that connects three or more amino acids
*   d. a covalent bond that connects two amino acids

47. Which of the following is considered one of the oldest kingdoms and

lacks a cell wall?

      a. kingdom protista

      b. kingdom plantae

*   c. kingdom archaebacteria

      d. kingdom eubacteria

There are six kingdoms, which consist of the following: kingdom protista, kingdom fungi, kingdom plantae, kingdom animalia, kingdom archaebacteria and kingdom eubacteria. Kingdom protista is made of mostly one-celled, eukaryotic organisms. Examples include heterophic and photosynthetic organisms. Kingdom plantae are eukaryotic organisms like plants and trees; they are multicellular. Kingdom archaebacteria do not have a peptidoglycan cell wall, and examples include thermophiles. Kingdom archaebacteria are prokaryotes. Kingdom eubacteria are also prokaryotes and have a cell wall made of peptidoglycan. Cyanobacteria are examples of kingdom eubacteria.

48. Which of the following is usually not true of eukaryotic organisms?

      a. multicellularity

      b. sexual reproduction

*   c. lack of a nucleus

      d. Algae are types of eukaryotes.

49. Which of the following carries hereditary information that is a kind of nucleic acid?

      a. glucose

      b. phenylalanine

*   c. DNA

      d. protease

Nucleic acids are macromolecules in cells that have the following parts in common: 5 carbon sugar, phosphate group and nitrogen base. DNA or deoxyribonucleic acid is made of the five carbon sugar deoxyribose, phosphate group and a nitrogen base. DNA is a double helix model. The basic parts of nucleic acids are nucleotides. There are bases in these nucleotides that are of two types: purines and pyrimidines. Adenine and guanine are purines, and cytosine and thymine are pyrimidines. Hydrogen bonds are between one purine and one pyrimidine. Referring to DNA and its paired bases, cytosine, a pyrimidine, only bonds with guanine, a purine. Thymine, a pyrimidine, bonds with adenine, a purine.

Structure and Function of the Cell & Energy Pathways

1. Which is the largest organelle in the eukaryotic cell?
    a. lysosome
*   b. nucleus
    c. mitochondria
    d. ribosome

2. DNA in the eukaryotic cell is found in the _____. In the prokaryotic cell, DNA is found in the _____.
*   a. nucleus, nucleoid
    b. cytoplasm, cytoplasm
    c. ribosome, golgi apparatus
    d. lysosome, ribosome

The nucleus is the largest structure of the cell; it contains genes. There is a nuclear envelope around the nucleus that partitions it from the cytoplasm. The cytoplasm contains all the rest of the cell's contents. The cytoplasm is found between the nucleus and the plasma membrane. Ribosome is a cell organelle where protein synthesis occurs. The golgi complex apparatus brings lipids and proteins to the plasma membrane. First, the protein is made in the ribosome and then stored in the rough endoplasmic reticulum. Later, the protein is brought to the golgi complex. The golgi complex can be thought as the Postal Service of the cell. This is because it packages proteins and delivers it to the outside of the cell or to other parts within the cell. Lysosomes are also cell organelles that use digestive enzymes to break down molecules and types of bacteria. Lysosomes are produced in the golgi complex.

3. Which of the following organelles digests bacteria through the process of phagocytosis?
    a. mitochondria
    b. nucleus
    c. ribosome
*   d. lysosome

4. Which is not a structure of the nucleus?
    a. chromatin
    b. nucleolus

        c. nuclear envelope
*   d. cytoskeleton
There is a nuclear envelope around the nucleus, which separates it from the cytoplasm. The nuclear envelope consists of two phospholipid membranes. The nucleus also has nuclear pores, which serve as an entrance to permit certain proteins to enter the nucleus. Inside the nucleus is the nucleolus where ribosomes are made. Chromatin is a thread-like material consisting of DNA and other proteins in the cell of the eukaryote. The cytoskeleton is composed of an arrangement of filament proteins found in the cytoplasm. The cytoskeleton is in charge of moving organelles and substances in the cell.

5. Chromatin fiber looks like beads or little marbles on a string. Each bead is made up of _____, and the entire bead is called a _____.
*   a. histones, nucleosome
        b. RNA, histone
        c. nucleosomes, double helix
        d. DNA, histone

6. Which is not a structure of the cytoskeleton?
        a. microtubules
        b. microfilaments
        c. intermediate filaments
*   d. mitochondria
The cytoskeleton is composed of filament proteins that help in the movement of organelles, chemicals and phagocytes in the cell. The cytoskeleton is composed of intermediate filaments, which aid in the giving and maintaining of cell shape. Microtubules aid in giving shape to the cell and in the production of flagella and cilia. Microtubules consist of a protein called tubulin. Microfilaments help in maintaining the shape of the cell. In skeletal muscle, actin and myosin are two types of microfilaments that contract muscles. Microfilaments are proteins. Intermediate filaments are quite strong and durable, and are also composed of proteins. Vimentin is the most common kind of intermediate filament and helps to preserve the structure of the cell.

7. Which of the following is not a type of intermediate filament?
        a. neurofilament

b. keratin
* c. myosin
d. vimentin

8. _____ are organelles that serve as the powerhouses of the cell; they perform the function of _____.
a. Lysosomes, phagocytosis
b. Ribosomes, protein synthesis
c. Cytoskeleton, support
* d. Mitochondria, cellular respiration

Mitochondria contain cristae, which provide added surface area to this organelle. Lysosomes are organelles that serve as part of the cell for digestion. Lysosomes get rid of used up organelles and old blood cells in the human body. Lysosomes can break down the following substances: proteins, lipids, carbohydrates and nucleic acids.

9. Secondary lysosomes have a pH that is _____; it is considered active and can break down other organelles. It uses hydrolytic enzymes in the process of _____.
* a. acidic, phagocytosis
b. acidic, cellular respiration
c. basic, phagocytosis
d. basic, cellular respiration

10. Which of the following generates ATP that is used for energy by the body?
a. cytoskeleton
* b. mitochondria
c. lysosomes
d. golgi apparatus

The golgi apparatus brings lipids and proteins to the plasma membrane. The cytoskeleton has a function of protection and movement of the cell. Lysosomes are the digestive organelles of the cell.

11. Which of the following is not a type of organelle?
a. mitochondria
b. ribosome
* c. cytoplasm

     d. nucleus

12. A cellular projection that is usually long and is the tail of a human sperm cell is _____.
     a. cilia
*   b. flagella
     c. centriole
     d. ribosome

Cilia, hair-like projections in eukaryotes, help in the movement of substances along the cell surface. A flagella aids in the movement of the whole cell. Centrioles are paired organelles in eukaryotes that are found near the nucleus. Centrioles are found in a centrosome. Each centriole is made of nine clusters of microtubule triplets. Microtubules are responsible for giving flagella and cilia their ability of movement.

13. Chlorophyll is a green pigment of plants found in _____.
*   a. chloroplasts
     b. centrioles
     c. mitochondria
     d. centrosomes

14. Which of the following is not a structure of a chloroplast?
     a. leucoplasts
     b. grana
*   c. mitochondria
     d. thylakoids

Amyloplasts, which are types of leucoplasts, serve as storage places for amylose. Chloroplasts become leucoplasts when there is a lack of sunlight. Thylakoids are coin-like spheres that contain chlorophyll; they are found in chloroplasts. Each stack of these thylakoids is called grana. Chloroplasts use energy from light to make its own food. This is done by the process of photosynthesis. Mitochondria are responsible for cellular respiration in the eukaryotic cell.

15. Which of the following organelles together are known as plastids?
*   a. amyloplast, chloroplast, leucoplast
     b. centrioles, centrosome, microtubules
     c. grana, thylakoids, stroma

d. grana, thylakoids, chloroplast

16. The inner membrane of a _____ can be described as a series of folds called _____.
    a. lysosome, cytoskeleton
*   b. mitochondrion, cristae
    c. nucleus, cristae
    d. centrosome, centrioles

Mitochondria are places where ATP is made for energy. Each mitochondrion has an outer membrane and an inner membrane. The inner membrane consists of a series of folds called cristae. The folds of cristae are important because it gives this organelle greater surface area to carry out the function of cellular respiration. Mitochondria break down glucose to make ATP with the use of oxygen. In this way, the body can have energy. The area of the mitochondria that is surrounded by the inner membrane and cristae is called the matrix. In other words, the spaces inside the cristae are where the matrix is located.

17. Which is between the outer and inner membranes of mitochondria?
    a. matrix
    b. cristae
    c. cisterns
*   d. intermembrane space

18. Animal cells take in water by the process of:
*   a. pinocytosis
    b. phagocytosis
    c. receptor-mediated endocytosis
    d. photosynthesis

Phagocytosis, pinocytosis and receptor-mediated endocytosis occur in the lysosome. In phagocytosis, pseudopods surround substances or organisms to eat. Pseudopods can be compared to lips that ingest food. When food is totally engulfed by pseudopods, the material that is surrounded is referred to as a phagosome. Pinocytosis doesn't need pseudopods. The cell just indents and lets liquid come in. Pinocytosis can occur in various cells of the body. In receptor-mediated endocytosis only specific substances are allowed into the cell. Cells that do this have indentions and are encrusted with clathrin, a protein.

19. _____ gets rid of hydrogen atoms in specific toxins in order to oxidize these certain poisons. In this way, it helps in detoxification of poisons in the human body.
    a. Golgi complex
*   b. Peroxisome
    c. Centrosome
    d. Ribosome

20. Photosynthesis occurs in which of the following?
    a. mitochondria
    b. ribosome
*   c. chloroplast
    d. lysosome

Chloroplasts are made of an outer and inner membrane. Inside the inner membrane are thylakoids that contain chlorophyll. A bunch of thylakoids is called grana. Ribosome is a small organelle that consists of a small subunit and a large subunit. Ribosomes contain ribosomal RNA. DNA in the nucleolus produces ribosomal RNA (rRNA).

21. Which organelle is responsible for lipid synthesis?
    a. rough endoplasmic reticulum
*   b. smooth endoplasmic reticulum
    c. mitochondria
    d. peroxisomes

22. Which organelle with the help of ribosomes aids in the synthesis of proteins?
    a. mitochondria
    b. smooth endoplasmic reticulum
*   c. rough endoplasmic reticulum
    d. golgi complex

The endoplasmic reticulum appears as folds of membranes. The portion of the endoplasmic reticulum with an abundant source of ribosomes is referred to as rough endoplasmic reticulum. The portion where there are no ribosomes is called the smooth endoplasmic reticulum. The golgi complex consists of 5-7 flattened folds. These flattened folds are called cisternae and are also known as cisterns. Proteins from the rough endoplasmic reticulum are transported to the cisternae. In the cisternae, they are put into packages

called secretory vesicles. These secretory vesicles will leave into the external part of the cell.

23. Entry into the cisternae of the golgi complex is by way of the _____. The exit of the cisternae in the golgi complex is by way of the _____.
      a. trans face, cis face
      b. transport vesicle, cis face
      c. cis face, rough endoplasmic reticulum
*   d. cis face, trans face

24. Which of the following organelles are formed in the golgi complex?
      a. cytoplasm
*   b. lysosome
      c. smooth endoplasmic reticulum
      d. nucleus
The cytoplasm is found between the exterior of the nucleus and interior to the plasma membrane of the cell. The nucleus is the control center of the cell. It controls the activities of the cell.

25. _____ and _____ are part of the theory of endosymbiosis. This theory states that these two structures have evolved when eukaryotes, from millions of years ago, engulfed bacteria cells.
*   a. Mitochondria and chloroplasts
      b. Centrioles and centrosomes
      c. Ribosomes and golgi complex
      d. Golgi complex and centrosome

26. Which of the following forms the foundation of the cell membrane?
*   a. phospholipid
      b. protein
      c. glucose
      d. cytoplasm
Phospholipid has one hydrophilic polar head and two hydrophobic tails. A phospholipid consists of glycerol attached to two fatty acids and a phosphorylated alcohol. The tails of the phospholipid, which are hydrophobic, act as barriers to water.

27. How does substances, such as glucose, cross the lipid bilayer of the cell membrane?

   a. by way of glucose floating on the bilayer of the cell membrane
*   b. by way of proteins floating on the bilayer of the cell membrane
   c. by way of nucleic acids floating on the bilayer of the cell membrane
   d. by way of ribosomes floating on the bilayer of the cell membrane

28. Referring to the structure of phospholipids suspended in the liquid bilayer of a cell membrane, the _____ of one phospholipid are attracted to the _____ of the other phospholipid.

   a. head, head
*   b. tails, tails
   c. head, tails
   d. tails, head

Membranes that are made up of a lipid bilayer of phospholipids usually surround cells. This phospholipid bilayer appears as one phospholipid head sticking outward towards the extracellular fluid and the other phospholipid head sticking outwards towards the inside, watery portion of the cell. The tails of the phospholipid closest to the outer edge of the cell line up with the tails of the phospholipid closest to the interior of the cell.

29. Which of the following is not a membrane lipid?

   a. glycolipid
   b. cholesterol
   c. phospholipid
*   d. anchors

30. According to the fluid mosaic model, which of the following attaches to anchoring proteins?

*   a. spectrin
   b. nucleus
   c. nucleolus
   d. golgi complex

The fluid mosaic model refers to proteins floating in fluid-like lipids. The plasma membrane is a good example of the fluid mosaic model. In this fluid mosaic model, you can see proteins anchored into the fluid-like lipids. Cytoskeleton anchors serve as a connection between filaments of the

cytoskeleton and the plasma membrane.

31. Which is a membrane protein that has configurations of beta-pleated sheets and form pores in the cell membrane?
        a. spectrin
*   b. porin
        c. phospholipid
        d. glycolipids

32. _____ and _____ are considered cell identity markers.
*   a. Glycoprotein, glycolipids
        b. Spectrin, porin
        c. Phospholipid, spectrins
        d. Clathrins, spectrins

Glycoprotein and glycolipids are considered cell identity markers or some sources refer to them as cell surface identity markers. Glycoproteins are composed of sugar and protein. Cell identity markers serve to tell your body the difference between your cells and someone else's cells. They recognize one's own cells. Each type of cell has specific shapes to glycoproteins and glycolipids. For example, blood types such as Type A and Type B are two different structures. They are not compatible. In this way, if a person with Type A blood was to get a transfusion of Type B blood, the cell identity markers would see this as foreign tissue. It would reject the transfusion. Clathrins are proteins that aid in the anchoring of specific proteins to certain places on the cell membrane.

33. The random movement of molecules from an area of high concentration to an area of lower concentration without the use of a carrier protein is referred to as:
        a. turgor pressure
        b. osmosis
*   c. diffusion
        d. facilitated diffusion

34. The movement of water from an area of high concentration to an area of lower concentration in a cell membrane is most closely defined as:
*   a. osmosis
        b. facilitated diffusion

c. turgor pressure

d. sodium-potassium pump

Water is diffused through the cell membrane by passing through the lipid molecules. Facilitated diffusion is the movement of substances, like glucose and amino acids, through the plasma membrane with the help of carrier proteins. For example, glucose concentration is higher on the outside of the cell. However, glucose concentration is lower on the inside of the cell. Facilitated diffusion lets the glucose concentration from the outside of the cell go to the area of lower concentration to the inside of the cell. Turgor pressure is also known as internal hydrostatic pressure. This occurs in plants. Sodium-potassium pump relates to sodium and potassium ions in animal cells. Sodium concentration is usually higher in the extracellular fluid and potassium concentration is higher in the cell itself. This pump only performs when ATP and sodium is in the extracellular fluid and potassium is in the cell. The sodium-potassium pump is actually a protein in the cell membrane that aids in the transport of sodium ions and potassium across this membrane. In this way, sodium can come into the cell and potassium can come out of the cell. However, this only happens by the sodium-potassium pump changing shape.

35. What happens directly after sodium ions bind with the sodium-potassium pump protein?

*   a. ATP phosphorylates the sodium-potassium pump protein.

       b. Sodium ions go into the cell.

       c. Potassium ions move out of the cell.

       d. ATP changes to DNA.

36. Relating to the sodium-potassium pump, after ATP phosphorylates the sodium-potassium pump, which will not occur next?

       a. ADP is formed.

       b. The pump protein changes shape.

*   c. Potassium ions leave the cell.

       d. Sodium ions leave the cell.

ADP is formed after ATP (adenosine triphosphate) is broken down into ADP (adenosine diphosphate) and a single phosphate. This single phosphate binds to the sodium-potassium pump, which is a protein. This adding a phosphate to the sodium-potassium pump is called phosphorylation. Once the sodium-potassium pump is phosphorylated, it changes its shape so three

sodium ions can leave to the outside of the cell. Since the sodium-potassium pump has changed its shape, two potassium ions now can bind to this pump. The binding of potassium ions with the sodium-potassium pump results in a dephosphorylation of this pump. In other words, the phosphate group is taken away from the sodium-potassium pump. This makes this pump return to its original shape and brings back the two potassium ions into the cell. The phosphate group is then added to the ADP (Adenosine diphosphate) producing ATP. ATP gives energy to animal cells.

37. Which of the following does not occur regarding the sodium-potassium pump?

    a. After a phosphate is removed from the sodium-potassium pump, ATP is formed.

    b. Sodium ions bind to the sodium-potassium pump.

    c. The phosphorylation of the sodium-potassium pump causes it to change its shape and causes ATP to change into ADP and a single phosphate.

*   d. Dephosphorylation of the protein pump causes it to change its shape so sodium ions can enter into the cell.

38. Which of the following gives the energy to the process of active transport across cell membranes?

*   a. ATP

    b. DNA

    c. fatty acid

    d. amino acid

Adenosine triphosphate (ATP) is what provides energy to animal cells. Active transport is the movement of specific substances through the cell membrane; the ATP makes this process possible. The cell membrane is also known as the plasma membrane. Active transport is the transfer of a substance across or through a cell membrane from a region of lower concentration to a region of a higher concentration. In other words, active transport transfers substances up its concentration gradient. Osmosis, diffusion and facilitated diffusion transfer substances down the concentration gradient. In other words, osmosis, diffusion and facilitated diffusion move substances from an area of higher concentration to an area of lower concentration.

39.   The sodium-potassium pump performs its function by way of
_____.
        a. osmosis
        b. facilitated diffusion
*   c. active transport
        d. endocytosis

40. The ingestion of solid material by the plasma membrane is called:
        a. exocytosis
*   b. phagocytosis
        c. pinocytosis
        d. chemiosmosis
Exocytosis is the release of substances from secretory vesicles at the outer
surface of the cell membrane. Exocytosis is used by animal cells to secrete
proteins, such as enzymes, and to secrete hormones. Pinocytosis and
phagocytosis are types of endocytosis. Endocytosis is the general process
of how the plasma membrane surrounds materials by expanding outward
and engulfing them. If the substance that the plasma membrane engulfs is
liquid, it is called pinocytosis. If the substance that the plasma membrane
engulfs is solid, it's phagocytosis. Chemiosmosis is involved with the
proton pump. Chemiosmosis helps with producing ATP from food in animal
cells.

41. _____ occurs in eukaryotic cells; indentations can be found
on the outside of the plasma membrane. _____ is a protein that is found
on the intracellular portion of the indentation. It is observed as a thin
covering.
*   a. Receptor-mediated endocytosis, Clathrin
        b. Pinocytosis, Spectrin
        c. Phagocytosis, Spectrin
        d. Chemiosmosis, Clathrin

42. According to cotransport across a cotransport channel, sodium ions go
back into the cell _____ its concentration gradient, and a substance, like
glucose, goes _____ its concentration gradient. This is another benefit
of the sodium-potassium pump.
        a. down, down
        b. up, up

        c. up, down
*   d. down, up

Amino acids and sugars benefit by cotransport. Sodium ions help to transport sugars or amino acids into the cell. Although sugars and amino acids are going up its concentration gradient into the cell, the strength of the force of sodium ion going down its concentration is much greater. Because the force of sodium ions going down the concentration gradient into the cell is much greater than force of the concentration of sugars and amino acids go against its concentration, the sodium ions carry in these molecules into the cell.

43. Which is a type of active transport through the cell membrane that helps in the generation of ATP?
        a. diffusion
        b. osmosis
*   c. proton pump
        d. facilitated diffusion

44. Which of the following is a type of cell junction?
        a. communicating
        b. tight
        c. anchoring
*   d. all of the above
Tight junctions are found between two cell membranes. Tight junctions connect each cell membrane together; they form very tight seals. Tight junctions prevent fluid and other molecules from passing between cells. Tight junctions are found in certain organs like the urinary bladder and the stomach. Anchoring junctions are found in cells that undergo stretching and physical stress like the skin. Adherens junctions, hemidesmosomes and desmosomes are types of anchoring junctions. Communicating junctions are ways that cells communicate with each other. The gap junction is a type of communication junction in animal cells. Communicating junctions serve to communicate signals to other cells in the form of action potentials or chemically. Action potentials are in the form of electrical signals. These action potentials are found abundant in the nerve cells of animals. In plants, cells communicate with each other by the use of plasmodesmata. These are openings of tiny, little tunnels that connect each plant cell's cytoplasm.

45. _____ acts as a tunnel between two animal cells. It allows for the transfer of ions between cells. In addition, the cytoplasm of both cells connects.
      a. Desmosome
      b. Tight junction
*   c. Connexon
      d. Cadherin

46. Which of the following anchors the epithelial cells to the basement membrane?
      a. desmosome
*   b. hemidesmosome
      c. tight junction
      d. gap junction
Desmosomes are anchoring junctions that connect cells together. The protein cadherin is found between the two plasma membranes of adjacent cells in a desmosome. Hemidesmosomes are half desmosomes, but they are found on the basement membrane.

47. Which kind of enzyme do most enzymic receptors represent?
*   a. protein kinase
      b. polymerase
      c. hydrolytic enyzme
      d. protease

48. Which of the following is not a type of cell surface receptor?
      a. G protein-linked receptors
      b. chemically gated ion channels
*   c. desmosomes
      d. enzymic receptors
Cell surface receptors aid in the transmission of bringing an extracellular signal to the intracellular portion of the cell. This brings on an altering inside to the cytoplasm of the cell. Chemically gated ion channels are made of proteins. Ions go through these channels or passageways to inside the cell. The gates open when a chemical called a neurotransmitter binds to the transmembrane proteins of this chemically gated ion channel. When the gates open, ions can come into the cell like sodium, calcium and potassium as examples. The ions come into the cell through a pore, which is between

the two transmembrane proteins of this chemically gated ion channel. The chemically gated ion channel is located on the plasma membrane. Most enzymic receptors are protein kinases that insert phosphate molecules to proteins. Enzymes are activated when a signal binds with the enzymatic receptor. These enzymatic receptors are composed of proteins. The region of the enzymatic receptor that binds with the signal is located on the extracellular part of the cell. The area of the enzymatic receptor that is activated is found in the cytoplasm. G protein-linked receptors are cell surface receptors that are linked with a guanosine triphosphate binding protein (G Protein). The G-protein linked receptor is found in the plasma membrane; it is a polypeptide chain. This chain makes a channel through the plasma membrane.

49. Which of the following is considered the communicating junction between plant cells?
     a. tight junctions
     b. gap junctions
     c. adherens junctions
*   d. plasmodesmata

50. A cell that sends signals to itself and connects with certain receptors is an example of:
     a. paracrine signaling
*   b. autocrine signaling
     c. endocrine signaling
     d. synaptic signaling
Endocrine signaling involves signals remaining in the extracellular fluid. In other words, the signal is released from a cell and enters the extracellular fluid and travels to other parts of the body by way of the circulatory system. This distant signal is referred to as hormones. Paracrine signaling involves signals having only restricted effects on nearby cells. Synaptic signaling involves the use of signals in the form of neurotransmitters. This is done by a nearby neuron secreting these neurotransmitters, which are transmitted to nerve cells called neurons.

51. Adherens junctions are types of _____, which link cells to one another and connect to microfilaments of the cytoskeleton.
*   a. anchoring junctions

b. communicating junctions

c. desmosomes

d. tight junctions

52. Which of the following is a cell surface marker that helps to identify cells from self and non-self?

    a. tight junction

*   b. major histocompatibility complex protein

    c. plasmodesmata

    d. desmosomes

Major histocompatibility complex protein or MHC protein is involved with the immune system of the body. MHC proteins help T cells and B cells help to recognize antigens or foreign substances in the body. These foreign substances in the body can be in the form of viruses, bacteria and organ transplants. T cells and B cells are types of antibodies that fight these foreign substances or antigens.

53. Reduction occurs when an atom _____ an electron, and oxidation occurs when an atom _____ an electron.

    a. gains, gains

    b. loses, loses

*   c. gains, loses

    d. loses, gains

54. The disorder of a system called _____ is most consistent with the _____ law of thermodynamics.

    a. entropy, first

*   b. entropy, second

    c. activation energy, first

    d. activation energy, second

The second law of thermodynamics is about the altering of potential energy into heat or unsystematic motion. In other words, matter tends to stay in an unordered form. The first law of thermodynamics says that the quantity of energy in the universe stays the same. In addition, energy is neither made nor destroyed; it just changes form. Activation energy is a specific quantity of energy needed to cause a chemical reaction.

55. According to the first and second laws of thermodynamics, energy in

the universe _____ and entropy is _____.

    a. remains the same, decreasing

*   b. remains the same, increasing

    c. is decreasing, decreasing

    d. is decreasing, increasing

56. Which is the process of decreasing activation energy so a chemical reaction can occur sooner and more frequently?

    a. thermodynamics

    b. oxidation

    c. reduction

*   d. catalysis

Thermodynamics is the study of energy. Oxidation occurs when an atom loses an electron; reduction occurs when an atom gains an electron. Since enzymes perform catalysis, they are known as catalysts. Remember most enzymes are proteins. Each enzyme usually has an indentation that is called an active site. The substrate binds with the enzyme at the active site. Thus, this forms the enzyme-substrate complex. However, the enzyme changes its own shape slightly in order to get a better hold of the substrate. This occurs when the substrate enters the active site of the enzyme.

57. Which of the following best defines a coenzyme?

    a. a protein molecule that increases the activation energy

*   b. non-protein organic molecules that are cofactors

    c. non-protein organic molecules that are substrates

    d. a protein molecule that decreases the activation energy

58. The majority of enzymes in the human body function best at _____°C. This point is referred to as _____. If the temperature goes higher than this, enzymes start to denature.

*   a. 35-40, temperature optimum

    b. 35-40, pH optimum

    c. 70-75, temperature optimum

    d. 60-65, activation energy

The pH optimum for most enzymes in the human body is between 6 and 8.

59. According to the Gibb's free energy equation, ΔG is represented as which of the following?

a. entropy
b. enthalpy
c. temperature
*   d. change in free energy

60. The _____ is the place where almost all noncompetitive inhibitors bind with an enzyme. In this way, the activity of the enzyme can be shut down.
*   a. allosteric site
b. active site
c. enzyme-substrate complex
d. cofactor

Allosteric inhibitor (noncompetitive inhibitor) is the substance that binds to the allosteric site on the enzyme. In this way, these allosteric inhibitors change the shape of the enzyme so the enzyme cannot bind with the substrate. Competitive inhibitor actual binds to the site on the enzyme where the substrate would bind. This blocks the substrate from binding with the enzyme at the active site. Cofactors help enzymes function. For example, certain ions, which are cofactors are observed in the active sites of enzymes. The enzyme-substrate complex is simply the binding of a substrate with an enzyme at the active site.

61. According to exergonic reactions, which of the following is not a true statement?
a. The reactant has more energy than the product.
b. Energy is released as heat.
*   c. ΔG is positive.
d. Exergonic reactions do not need added energy to occur. However, activation energy is needed to start these reactions.

62. The active part of NAD+ is _____. The part that holds the shape of NAD+ is called _____.
*   a. NMP, AMP
b. AMP, NMP
c. ADP, ATP
d. ATP, ADP

NAD+ or nicotinamide adenine dinucleotide is a coenzyme. NAD+ is made of two nucleotides.   One nucleotide is called NMP or nictinamide

monophosphate and the other is called AMP or adenine monophosphate. Coenzymes accept electrons. They then transfer these electrons to enzymes so more chemical reactions can take place.

63. The building up of molecules is referred to as _____. The breaking down of molecules is called _____ and digestion is an example.
        a. catabolism, anabolism
*   b. anabolism, catabolism
        c. glycolysis, degradation
        d. degradation, glycolysis

64. Which is not part of the structure of an ATP molecule?
        a. triphosphate group
        b. AMP group
        c. ribose
*   d. NAD+
ATP is the main energy supply of all cells. ATP is basically one ribose molecule, three phosphate groups and adenine, a nitrogenous base. However, AMP group is basically adenine and a phosphate connected by the ribose group. This AMP group is found both in ATP and NAD+. Remember that ribose is a five-carbon sugar.

65. Energy is mostly stored in the _____ of an ATP molecule.
When this portion of the ATP molecule is cleaved, _____ is produced.
        a. adenine, NAD+
*   b. triphosphate group, ADP
        c. ribose, ADP
        d. adenine, NAD+

66. The phosphate bonds between phosphate molecules are _____ in ATP. These bonds _____ one another and are _____ charged.
        a. stable, attract, positively
        b. unstable, attract, positively
        c. stable, repel, negatively
*   d. unstable, repel, negatively
Endergonic reactions in cells occur because of ATP. Endergonic reactions need energy to occur; ATP supplies this energy.

67. The sum total of all catabolic and anabolic processes in the cells of an animal or other life form is referred to as:
*   a. metabolism
       b. entropy
       c. catalysis
       d. activation energy

68. Which is the process of acquiring nitrogen atoms from N_2 gas, which some types of bacteria still use today?
       a. glycolysis
*   b. nitrogen fixation
       c. cellular respiration
       d. photosynthesis
Glycolysis is the breakdown of a molecule of glucose into two pyruvic acid molecules. The whole process of glycolysis results in the creation of 2 molecules of ATP. Nitrogen fixation occurs in places where there is no oxygen. Certain bacteria can live without oxygen; they are called anaerobic bacteria. Cellular respiration is referred to as the oxidation of glucose or other organic molecules to make ATP.

69. All of the following are types of autotrophs except:
       a. plants
       b. minority of bacteria
       c. algae
*   d. animals

70. Organisms that use inorganic molecules to make its own organic molecules as its food for growth are referred to as _____.
       a. heterotrophs
*   b. autotrophs
       c. animals
       d. fungi
Autotrophs also use the process of photosynthesis, which changes light energy into chemical energy. Plants, algae, some protists and bacteria are examples of autotrophs. Heterotrophs cannot get energy from photosynthesis. Instead, they eat organic molecules such as the ones found in animals and plants to get energy. The majority of bacteria, fungi and animals are examples of heterotrophs.

71. Which of the following is a pathway used in aerobic respiration to produce energy from glucose?
    a. pyruvate oxidation
    b. glycolysis and Krebs cycle only
    c. electron transport chain
*   d. all of the above

72. There are _____ reactions that occur in glycolysis. With the help of two molecules of ATP, phosphorylation of glucose occurs in step _____ of glycolysis.
*   a. 10, 1
    b. 8, 1
    c. 12, 3
    d. 10, 4

The first reaction (step 1) of glycolysis involves the phosphorylation of glucose by ATP. In other words, an additional phosphate group is added to glucose. The ATP then becomes ADP. With the help of the enzyme, hexokinase, in this step one of glycolysis, glucose now becomes glucose 6-phosphate. The second reaction (step 2) of glycolysis occurs when glucose 6-phosphate with the help of phosphoglucoisomerase becomes fructose 6-phosphate. The third reaction (step 3) of glycolysis occurs when fructose 6-phosphate is phosphorylated by ATP. Because of this addition of a phosphate group to fructose 6-phosphate, ATP turns into ADP. Also, fructose 6-phosphate with the help of phosphofructokinase now becomes fructose 1, 6-bisphosphate.

73. Step 3 of glycolysis involves the phosphorylation of _____ by _____ and the result is fructose 1, 6 bisphosphate with the help of the enzyme phosphofructokinase.
    a. glucose, ATP
    b. glucose 6-phosphate, ADP
    c. fructose 6-phosphate, ATP
*   d. fructose 6-phosphate, ADP

74. In step four of glycolysis, fructose 1, 6-bisphosphate is split into dihydroxyacetone phosphate and _____.
    a. glucose
*   b. glyceraldehyde 3-phosphate

c. fructose 6-phosphate

d. glucose 6-phosphate

Fructose 1, 6-bisphosphate with the help of the enzyme, aldolase, becomes the three carbon molecule dihydroxacetone phosphate and the three carbon molecule glyceraldehyde 3-phosphate. In step 5 of glycolysis, dihydroxacetone phosphate is changed into glyceraldehyde 3-phosphate. So now there are two molecules of glyceraldehyde 3-phosphate. In step 6, two electrons are moved from one molecule glyceraldehyde 3-phosphate to NAD+. This makes NADH. Also, two electrons are moved from the other molecule of glyceraldehyde 3-phosphate to NAD+. This also makes NADH. Then both molecules are phosphorylated; two molecules of 1,3-bisphosphoglycerate form. In step 7, each molecule of 1,3-bisphosphoglycerate loses a phosphate group by the help of two ADP. Because of this, two ATP are made, and each molecule of 1, 3-bisphosphoglycerate becomes 3-phosphoglycerate. So now there are two molecules of 3-phosphoglycerate.

75. In step 7 of glycolysis, with the help of the enzyme, phosphoglycerokinase, and the removal of a phosphate group by ADP, 1, 3-bisphosphoglycerate becomes _____. Because of this _____ is produced.

* a. 3-phosphoglycerate, ATP

b. glucose 6-phosphate, ATP

c. fructose 1, 6-bisphosphate, NADH

d. fructose 6-phosphate, NADH

76. In step 8 of glycolysis, two molecules of 3-phosphoglycerate become _____ with the help of the enzyme phosphoglyceromutase.

a. glucose 6-phosphate

* b. 2-phosphoglycerate

c. glucose

d. 1, 3-bisphosphoglycerate

In step 9 of glycolysis, each molecule of 3-phosphoglycerate loses a molecule of water. With the help of enolase, an enzyme, each molecule of 3-phosphoglycerate becomes phosphoenolpyruvate. Finally, in step 10, ADP removes a phosphate group from each molecule of phosphoenolpyruvate. Because of this removal of a phosphate group, the ADP now becomes ATP. Thus, two ATP molecules are made. With the help

of pyruvate kinase, an enzyme, each molecule of phosphoenolpyruvate now becomes pyruvate.

77. There are _____ ATP molecules produced in glycolysis. However, there is only a net gain of _____ ATP molecules since two ATP molecules are used in steps 1 and 3 of glycolysis.
     a. 8, 4
     b. 4, 4
*   c. 4, 2
     d. 6, 2

78. When pyruvate is oxidized, _____ forms. This takes place in the _____.
*   a. acetyl-CoA, mitochondria
     b. acetyl-CoA, ribosome
     c. ATP, mitochondria
     d. ATP, ribosome
Pyruvate is oxidized in aerobic conditions. This occurs by one decarboxylation of pyruvate. In other words, one carbon of pyruvate is cut off with the help of pyruvate dehydrogenase and CO_2 is removed. The other two carbons with the help of NAD+ being reduced to NADH become an acetyl group. The acetyl group than combines with coenzyme A to form acetyl-CoA. The decarboxylation of pyruvate occurs in the matrix of the mitochondria.

79. When pyruvate is decarboxylated, _____ is reduced to NADH and _____ is lost. Thus, the end result is ultimately an acetyl group combining with coenzyme A (CoA) to form acetyl-CoA.
     a. ATP, ADP
     b. ADP, ATP
     c. CO_2, NAD+
*   d. NAD+, CO_2

80. The Krebs cycle consists of _____ reactions. _____ first enters this cycle.
     a. 10, Succinyl-CoA
     b. 8, Malate
*   c. 9, Acetyl-CoA

d. 9, Malate

The Krebs cycle is also referred to as the citric acid cycle. Most sources say that there are 9 reactions or steps in the Krebs cycle. However, a few sources state that the Krebs cycle is only 8 steps. Since most sources and most colleges and universities teach it as a 9 step process, this is the most reasonable approach. The Krebs cycle occurs in the mitochondria matrix. The first reaction or step 1 occurs with acetyl-CoA joining with oxaloacetate. This condensation reaction forms citrate (citric acid). This is done with the help of the enzyme citrate synthetase. CoA is just used to bring acetyl into the Krebs cycle. Remember the main purpose of the Krebs cycle is to make ATP. In step 2 and 3 of the Krebs cycle, a water molecule changes position on citrate and becomes isocitrate. This occurs with the help of enzyme, aconitase.

81. Steps 2 and 3 of the Krebs cycle are best represented as which type of reaction?
    a. condensation
    b. substrate-level phosphorylation
*   c. isomerization
    d. oxidation

82. Isocitrate is oxidized in step _____ of the Krebs cycle. _____ is produced in this step.
*   a. 4, α-Ketoglutarate
    b. 2, Citrate
    c. 3, Citrate
    d. 6, Succinate

In Step 4 of the Krebs cycle, isocitrate is oxidized. Also, NAD+ is reduced to NADH. Then isocitrate is decarboxylated; isocitrate losses a CO_2. So isocitrate now becomes α-ketoglutarate. In step 5, α-ketoglutarate is oxidized and decarboxylated with the help of α-ketoglutarate dehydrogenase, an enzyme. In other words, CO_2 group is taken off α-ketoglutarate and connects with coenzyme A, which forms succinyl-CoA. Also, NAD+ is reduced to NADH.

83. α-Ketoglutarate becomes _____ in _____ of the Krebs cycle with the help of the enzyme α-ketoglutarate dehydrogenase.
    a. malate, step 5

*   b. succinyl-CoA, step 5
     c. isocitrate, step 3
     d. fumarate, step 7

84. Succinyl-CoA becomes _____ in step 6 of the Krebs cycle. _____ is the enzyme involved in this reaction.
     a. isocitrate, Isocitrate dehydrogenase
     b. citrate, Citrate synthetase
*   c. succinate, Succinyl-CoA synthetase
     d. fumarate, Succinate dehydrogenase

In step 6 of the Krebs cycle, there is a couple reaction. Succinyl-CoA becomes succinate with the help of enzyme, succinyl-CoA synthetase. Also, in step 6, guanosine diphosphate (GDP) is phosphorylated to form guanosine triphosphate (GTP). GTP then can give one of its phosphate groups to ADP. Then ADP is converted to ATP. In step 7, succinate is oxidized to fumarate with the help of succinate dehydrogenase. Flavin adenine dinucleotide (FAD+) accepts the electrons from succinate. FAD+ is then reduced to $FADH_2$. $FADH_2$ gives electrons to the electron transport chain.

85. Step 6 of the Krebs cycle is best represented as which of the following reactions?
*   a. substrate-level phosphorylation
     b. condensation
     c. isomerization
     d. all of the above

86. In step 8 of the Krebs cycle, _____ is added to fumarate and with the help of fumarase _____ is created.
     a. phosphate group, malate
*   b. H_2O, malate
     c. $FADH_2$, succinate
     d. NAD+, α-ketoglutarate

In step 9, malate is oxidized with the help of malate dehydrogenase to form oxaloacetate. NAD+ is also reduced to NADH by the dehydrogenation of malate. Oxaloacetate can now connect with acetyl to start the Krebs cycle again.

87. Which of the following occurs at the beginning of the Krebs cycle?
       a. succinate is converted to fumarate
       b. isocitrate is converted to α-ketoglutarate
       c. succinate is oxidated
*   d. oxaloacetate combines with acetyl

88. Electrons from NADH and _____ go into the electron transport chain in order to ultimately produce ATP.
*   a. FADH$_2$
       b. isocitrate
       c. succinyl-CoA
       d. citrate

The process of ATP generation by facilitated diffusion is referred to as chemiosmosis. This occurs by the transfer of protons in the direction of the negatively charged mitochondria matrix. Remember that electrons pass through the electron transport chain from steps of high energy to steps of lower and lower energy. The electron transport chain involves carrier molecules that are located on the internal membrane of the mitochondria.

89. The first step in the electron transport chain involves the movement of high energy electrons from:
*   a. NADH and H+ to FMN
       b. Q to cyt b
       c. Fe to S
       d. FMNH$_2$ to FMN

90. Which of the following acts as a protein pump in the electron transport chain?
       a. citrate
       b. isocitrate
*   c. NADH dehydrogenase complex
       d. acetyl-CoA

NADH dehydrogenase complex consists of FMN (flavin mononucleotide) and iron-sulfur centers. The iron-sulfur centers are the places where the electrons can be transferred in NADH dehydrogenase. After the first step in the electron transport chain, proteins are then carried again by ubiquinones, which are non-protein electron carriers. From the ubiquinones, the electrons are carried to a complex called the cytochrome complex (bc$_1$

complex). The cytochrome complex and the NADH dehydrogenase complexes are considered proton pumps. From the cytochrome complex, the electrons go to the cytochrome oxidase complex. The cytochrome oxidase complex is the third type of proton pump in the electron transport chain. The electrons from NADH are put into NADH hydrogenase (high energy level) in the electron transport chain. The electrons from $FADH_2$ are put into the electron transport chain at a lower energy level; these electrons enter at the ubiquinone (lower energy level). The last electron acceptor in the electron transport chain is oxygen.

91. Which is not a type of proton pump found in the electron transport chain?
     a. cytochrome oxidase complex
     b. cytochrome complex
     c. NADH dehydrogenase complex
*   d. acetyl CoA

92. Protons are moved out of the matrix of the mitochondria into the intermembrane space by the _____.
     a. carrier acetyl CoA
*   b. energy released by the electron transport chain
     c. enzyme pyruvate kinase
     d. the enzyme hexokinase
Energy is released by the electron transport chain as a result of the transport of electrons through this chain. When the amount of protons in the intermembrane space is high enough, these protons go back into the mitochondrial matrix by the process of diffusion. The loss of protons in the mitochondrial matrix causes it (the matrix) to become negative in charge. The protons go back to mitochondrial spaces since positive charges are attracted to negative charges. When the protons come back into the mitochondrial space by way of proton channels, ATP is made inside the mitochondria.

93. What happens to the electrons in the electron transport chain after the protons are pumped out from the mitochondrial matrix and into the intramembrane space of the mitochondria?
*   a. they are accepted by oxygen to form water
     b. they are donated to FMN

c. they are accepted by enolase

d. they are donated to aldolase

94.   ATP that is produced without oxygen occurs in the process of _____.

      a. chemiosmosis

      b. aerobic respiration

*   c. fermentation

      d. photosynthesis

Photosynthesis and aerobic respiration use chemiosmosis as a process of producing ATP.

95. Glucose catabolism consists of which of the following?

      a. glycolysis

      b. Krebs cycle

      c. electron transport chain

*   d. all of the above

96. The oxidation of one molecule of glucose can produce approximately up to _____ ATP. _____ does not require oxygen.

      a. 15, Krebs cycle

      b. 10, Krebs cycle

      c. 36, Electron transport chain

*   d. 36, Glycolysis

Cellular respiration refers to the oxidation of glucose. Cellular respiration can be also known as oxidative respiration. Glucose is oxidized by glycolysis and aerobically oxidized by the Krebs cycle and the electron transport chain. Because of this oxidation of glucose, ATP is produced.

97. Fatty acids are oxidized by the process of _____, which occurs in the matrix of the mitochondria.

*   a. beta-oxidation

      b. glycolysis

      c. deamination

      d. osmosis

98. There are _____ carbons removed at a time from each fatty acid during beta-oxidation. These two carbons then connect with coenzyme A to

form _____.
    a. 5, acetyl-CoA
*   b. 2, acetyl-CoA
    c. 7, pyruvate
    d. 9, pyruvate

The acetyl CoA formed is needed for the Krebs cycle. A 16-carbon fatty acid can have each of its two carbon units to be oxidized. Each of the two carbon units connects with coenzyme A to form Acetyl-CoA. Each Acetyl-CoA then goes to the Krebs cycle. Oxidation of a fatty acid involves first beta-oxidation, then the Krebs cycle, and finally the electron transport chain. Oxidation of a fatty acid can also be expressed as cellular respiration of fatty acids.

99. An amino group (NH_2) is removed from an amino acid in an aerobic environment by the process of _____. These altered amino acids can then go into the _____.
    a. fermentation, electron transport chain
    b. Beta oxidation, electron transport chain
*   c. deamination, Krebs cycle
    d. fermentation, Krebs cycle

100. The process of photosynthesis takes place in the leaves of plants. Which is the first step or stage in photosynthesis?
    a. using energy from sunlight to manufacture ATP
*   b. acquire energy from the light of the sun
    c. use light-independent reactions
    d. synthesis of organic substances from carbon dioxide

The second step in photosynthesis involves using the energy from sunlight to manufacture ATP and to reduce NADP to NADPH. The first two steps or stages of photosynthesis are known as light-dependent reactions. The third stage of photosynthesis is the synthesis of organic substances from carbon dioxide. This is achieved by using ATP and NADPH in the manufacturing of organic molecules from carbon dioxide. The third stage of photosynthesis can occur in the lack or presence of light. Therefore, the third stage is known as light-independent reactions.

101. Which of the following is the third stage of photosynthesis?
*   a. synthesis of organic substances from carbon dioxide with the aid of

ATP and NADPH
b. getting sunlight from the sun
c. making ATP
d. all of the above

102. Chloroplasts contain _____, which are the locations where photosynthesis occurs.
    a. ATP
    b. RNA
    c. stroma
*   d. thylakoid membranes

Plants have leaves that have many cells which contain chloroplasts. These many cells form a thick layer, which is called the mesophyll. In the mesophyll are chloroplasts. Chloroplasts are composed of thylakoids. Inside the thylakoid membranes are the location for the reactions of photosynthesis. A stack of thylakoids is referred to as granum. Each granum is enclosed by stroma, which is a partial liquid material.

103. The photosystem can be most closely described as:
*   a. chlorophyll pigments that form a unit, which are located within the thylakoid membranes
    b. stroma
    c. the production of ATP
    d. the cell wall that is located near the mesophyll

104. Which of the following is the accessory light absorbing pigment in plants?
    a. chlorophyll *a*
*   b. chlorophyll *b*
    c. photon
    d. NADPH

Chlorophyll *a* is a primary photosynthetic pigment in plants and can change light energy to chemical energy. Light energy is absorbed in the form of photons. Chlorophyll *b* adds to the absorption of light with chlorophyll *a*. Chlorophyll *a* and chlorophyll *b* absorb red and violet-blue light. According to the absorption spectrum, these two pigments actually reflect green light. This is why plants appear green.

105.   Carotenoid is a pigment also used in photosynthesis; it absorbs
_____ light and reflects _____ light.
*   a. blue-green, orange and red
        b. orange and red, blue-green
        c. orange and red, violet-blue
        d. violet-blue, green

106.   Which of the following is considered a primary pigment and not an
accessory type that absorbs photons of light?
        a. carotenoid
*   b. chlorophyll *a*
        c. chlorophyll *b*
        d. all of the above
Carotenoids are accessory pigments that can secure energy from the light of
wavelengths that are not effectively absorbed by chlorophyll. Chlorophyll *a*
and chlorophyll *b* have chemical structures of a porphyrin head and a tail
composed of hydrocarbon. The porphyrin head is a ring structure. The
difference in the porphyrin head in chlorophyll *a* is that it has a CHO group
and chlorophyll *b* has a CH_3 group. The CHO group is an aldehyde and the
CH_3 group is a methyl group.

107.   Electrons in the porphyrin ring are excited by the absorbed _____
in _____ and _____.
        a. water, carotenoid, chlorophyll *a*
        b. water, carotenoid, chlorophyll *b*
*   c. photons, chlorophyll *a*, chlorophyll *b*
        d. photons, carotenoid, chlorophyll *b*

108.   According to the electromagnetic spectrum in regards to visible light,
which of the following is a true statement?
        a. Gamma rays have longer wavelengths than visible light.
*   b. A longer wavelength of visible light has less energy than a shorter
        wavelength of visible light.
        c. Radio waves have shorter wavelengths than visible light.
        d. A longer wavelength of visible light has more energy than a shorter
        wavelength of visible light.
The electromagnetic spectrum shows the different forms of electromagnetic
energy. Gamma rays have the shortest wavelength and the most energy in

the electromagnetic spectrum. The next in line is x-rays, which have a longer wavelength than gamma rays. After x-ray, UV light is the next wavelength and has a longer wavelength. The next slightly lower wavelength after UV light is visible light. Visible light is approximately between 400 and 750 nanometers (nm). Violet has the shortest wavelength while red has the longest. Violet visible light, which has a shorter wavelength, has more energy than red visible light. Red visible light has a longer wavelength. Radio waves and infrared light have longer wavelengths than visible light.

109. Violet light has a shorter wavelength and photons with _____ energy than _____.
*   a. more, red light
       b. less, red light
       c. less, radio waves
       d. more, gamma rays

110. Which of the following is part of the photosystem in chloroplasts?
       a. antenna complex
       b. reaction center
       c. chlorophyll *a* molecules
*   d. all of the above
Photons are acquired by the antenna complex. These photons are obtained by sunlight or the visible light. The antenna complex consists of molecules of chlorophyll that are connected on the thylakoid membrane by proteins. This antenna complex brings the acquired photons (light energy) to the reaction center. In the reaction center, there are chlorophyll *a* molecules that transfer the light energy away from this photosystem.

111. The reaction center in a photosystem can be best described as:
       a. antenna complex
       b. visible light
*   c. protein-pigment complex
       d. quinone

112. Excitation energy from the absorption of photons is transported to pigment molecule to pigment molecule until it (this excitation energy) goes to _____ in the _____.

*   a. chlorophyll *a*, reaction center
         b. carotenoid, reaction center
         c. ATP, antenna complex
         d. photons, absorption spectrum

The reaction center brings energy out of the photosystem. The light energy that is brought out of the photosystem will then be converted to chemical energy. Remember that when a pigment, such as chlorophyll b absorbs photons (light energy), the pigment's electrons go from a lower energy level to a higher energy level.

113. Which of the following occurs in photosystem II in plants?
*   a. photon from light expels a high-energy electron from photosystem II
         b. photon from light makes NADP
         c. photon from light makes NADPH
         d. photosystem I gets a photon of light

114. The reaction center pigment in photosystem II is referred to as _____. A proton is driven across the thylakoid membrane to help form _____.

*   a. P_{680}, ATP
         b. P_{700}, ATP
         c. P_{700}, NADP reductase
         d. P_{680}, NADP reductase

In photosystem II, photons of light drive electrons out of P_{680}. The symbol P_{680} refers to the pigments that absorb photons specifically at 680 nanometers. When an electron is removed from these pigments in the reaction center and goes out of photosystem II, the Z protein (enzyme) splits a molecule of water. This splitting of water causes electrons from the water to be put back in the reaction center. When two molecules of water are split, four electrons are removed, and oxygen is then discharged out into the atmosphere. The electrons must be put back in the reaction center of photosystem II since some were driven out by a photon. The electrons that were driven out help to bring protons, which were taken from water molecules, to the cytochrome electron carriers. Plastoquinone is the electron carrier that brings the excited electrons to the proton pump, called b_6-f complex. When the excited electrons arrive at the proton pump, protons are pumped inside the thylakoid. The protons then diffuse out of

the thylakoid by way of ATP synthase channels. When this happens, ADP gets phosphorylated to form ATP. Thus, the protons and ATP go into the stroma.

115. _____ is an electron carrier that carries electrons to the proton pump referred to as _____ after leaving photosystem II.
     a. Ferredoxin, b_6-f complex
     b. NADP, plastoquinone (Q)
     c. Plastoquinone, ferredoxin
*   d. Plastoquinone (Q), b_6-f complex

116. Which of the following is a protein that brings electrons from photosystem II from the protein pump to photosystem I?
     a. plastoquinone
*   b. plastocyanin
     c. Z protein
     d. ferredoxin
The excited electrons are carried by plastoquinone to the protein pump and then to plastocyanin (PC). Plastocyanin then brings the electrons to photosystem I.

117. Photosystem I absorbs photons and this causes electrons to be brought to _____. These electrons are then brought to NADP reductase. NADP is then reduced to _____.
*   a. ferredoxin, NADPH
     b. NADPH, ferredoxin
     c. plastoquinone, ATP
     d. ferredoxin, ATP

118. Which of the following is produced by noncyclic photophosphorylation?
     a. NADPH only
     b. ATP only
*   c. ATP, NADPH
     d. zinc
In noncyclic photophosphorylation, photosystem II occurs first to produce ATP. Then photosystem I produces NAPDH. Many types of plants can also use cyclic photophosphorylation in addition to noncyclic

photophosphorylation. When more ATP needs to be produced, many plants use cyclic photophosphorylation. Although NADPH is not produced, more ATP is made in cyclic photophosphorylation.

119. When noncyclic photophosphorylation changes to cyclic photophosphorylation, the excited electron leaves photosystem I and goes back to the _____ to make more _____.
*   a. b_6-f complex, ATP
     b. ferredoxin, NADP
     c. ferredoxin, NADPH
     d. b_6-f complex, NADPH

120. Ribulose 1, 5-bisphosphate (RuBP) combines with three molecules of carbon dioxide in the first step of the Calvin Cycle and with help of the enzyme, rubisco (ribulose bisphosphate carboxylase), form two molecules of _____.
     a. glyceraldehyde 3-phosphate
     b. 1,3-bisphosphoglycerate
     c. ribulose 1,5-bisphosphate
*   d. 3-phosphoglycerate

The ATP and NAPDH (reducing power) made in photosystem I and photosystem II are utilized to help fix carbon in the Calvin cycle. To fix carbon means to change CO_2 into carbohydrates. CO_2 is changed into carbohydrates in the form of sugars in the Calvin cycle. The Calvin cycle takes place in the stroma portion of the chloroplast.

121. In photosynthesis, ATP and NADPH are produced in _____ reactions, and the Calvin cycle usually occurs in _____ reactions.
     a. light-independent, light-dependent
*   b. light-dependent, light-independent
     c. light-independent, light-independent
     d. dark, light

122. In the Calvin cycle, 3-phosphoglycerate (PGA) with the aid of the enzyme, PGA kinase, produces six molecules of _____.
*   a. 1,3-bisphosphoglycerate
     b. glyceraldehyde 3-phosphate
     c. ribulose 1,5-bisphosphate (RuBP)

d. 3-phosphoglycerate

Six molecules of 3-phosphoglycerate with the help of PGA kinase become six molecules of 1,3-bisphosphoglycerate. ATP is needed in the Calvin cycle to make this energy pathway occur.

123. Six molecules of 1,3-bisphosphoglycerate with the help of the enzyme, G3P dehydrogenase, form six molecules of _____.
    a. PGA kinase
    b. 3-phosphoglycerate
    c. ribulose 1, 5-bisphosphate
*   d. glyceraldehyde 3-phosphate

124. One of the six molecules of glyceraldehyde 3-phosphate in the Calvin cycle can be converted to glucose 1-phosphate, which will eventually form _____. The other five molecules of glyceraldehyde 3-phosphate will then reform into _____.
    a. starch, carbon dioxide
*   b. starch, ribulose 1,5-bisphosphate (RuBP)
    c. glucose, carbon dioxide
    d. glucose, 1-3,bisphosphoglycerate

Ribulose 1,5-bisphosphate (RuBP) will eventually start the whole Calvin cycle again. The Calvin cycle will occur again as long as there is enough carbon dioxide available and energy in the form of ATP. Also, NADPH is also needed because it gives a supply of hydrogens and energized electrons that can connect them with carbon. In this way, in the Calvin cycle, 1,3-bisphosphoglycerate can form glyceraldehyde 3-phosphate.

125. When carbon dioxide from the atmosphere connects with 1,5-bisphosphate (RuBP) to produce phosphoglycerate molecules, this is most closely defined as:
*   a. C_3 photosynthesis
    b. glycolysis
    c. photosystem I
    d. photosystem II

126. Which is the correct order in the Calvin cycle from the beginning?
    a. glyceraldehyde 3-phosphate, 3-phosphoglycerate, ribulose 1,5-bisphosphate, 1,3-bisphosphate, glyceraldehyde 3-phosphate

b. 1,3-bisphosphate, 3-phosphoglycerate, glyceraldehyde 3-phosphate, ribulose 1,5 bisphosphate

* c. ribulose 1,5-bisphosphate, 3-phosphoglycerate, 1,3-bisphosphate, glyceraldehyde 3-phosphate, ribulose 1,5-bisphosphate

d. carbon dioxide, 3-phosphoglycerate, 1,3-bisphosphoglycerate, ribulose 1,5-bisphosphate, glyceraldehyde 3-phosphate

The Calvin cycle does not need light to occur. The Calvin cycle is known as a light-independent reaction or a dark reaction.

127. Which of the following is the process that changes the carbon fixed by photosynthesis back to carbon dioxide?

* a. photorespiration

b. glycolysis

c. chemiosmosis

d. carbon fixation

128. Which enzyme carries out carbon fixation and photorespiration?

a. PGA kinase

b. NADP reductase

c. ATP synthase

* d. ribulose bisphosphate carboxylase

NADP reductase is involved with noncyclic photophosphorylation. ATP synthase is involved with noncyclic and cyclic photophosphorylation. PGA kinase is an enzyme in the Krebs cycle that helps to change 3-phosphoglycerate (PGA) into 1,3-bisphosphoglycerate.

129. The starting substance in C_4 photosynthesis is referred to as _____, which contains four carbons.

* a. oxaloacetate

b. glyceraldehyde 3-phosphate

c. 1,3-bisphosphoglycerate

d. starch

130. Which of the following plants uses crassulacean acid metabolism (CAM) as the pathway for carbon fixation?

a. pineapples

b. cacti

c. succulent plants

*   d. all of the above

Crassulacean acid metabolism involves the stomata, which are the openings in the leaves of plants in water storing plants such as pineapples. These stomata take in carbon dioxide; they close during the day and open at night. The reason why they close during the day is to keep carbon dioxide inside the plant. Without the carbon dioxide present, the sugars in the pineapple could not be produced. Crassulacean acid metabolism occurs in plants that are in very hot climates, which are tropical or very dry.

131.   The amount of ATP used in C_4 photosynthesis to convert carbon dioxide to glucose is _____.
*   a. about double to the amount used in C_3 photosynthesis
       b. about equal to the amount used in C_3 photosynthesis
       c. about one-quarter to the amount used in C_3 photosynthesis
       d. about one-half to the amount used in C_3 photosynthesis

132.   Plants in warm climates utilize C_4 photosynthesis in order to store high amounts of CO_2 in the _____ of its leaves. In this way, this prevents a lot of the CO_2 from being lost to photorespiration.
       a. epidermis
*   b. bundle-sheath cells
       c. spongy parenchyma
       d. stoma

In plants that utilize C_3 photosynthesis, CO_2 is taken up by mesophyll cells. In order to perform C_4 photosynthesis, phosphoenolpyruvate is carboxylated to form oxaloacetate. Oxaloacetate is than changed into malate. Malate is then brought to the bundle-sheath cells. Malate is then decarboxylated in the bundle-sheath cells. This produces pyruvate and more importantly CO_2. This CO_2 will then enter the Calvin cycle.

Reproduction and Heredity

1. Which is not true of bacteria?
       a. They reproduce by binary fission.
       b. They lack a nucleus.
*   c. They reproduce by meiosis.
       d. They are prokaryotes.

2. Asexual reproduction of a bacterial cell duplicates its DNA and then divides into approximate equal portions. This occurs in the process of _____.
*   a. binary fission
        b. mitosis
        c. anaphase
        d. telophase

The complete genetic information in an organism is called its genome. Replication or copying of DNA (genetic information) in bacterium occurs in the replication origin. The two equal portions that form from binary fission are called daughter cells. Binary fission starts at the region where two genomes are connected to the plasma membrane of the bacteria cell. Anaphase and telophase are phases in mitosis.

3. Normal human cells have _____ pairs of chromosomes. This consists of _____ chromosomes in each human cell.
        a. 24, 48
*   b. 23, 46
        c. 32, 64
        d. 36, 72

4. Chromatin is found in _____; it is made up of protein and _____.
        a. eukaryotic chromosomes, cytoplasm
        b. prokaryotic chromosomes, RNA
        c. prokaryotic chromosomes, DNA
*   d. eukaryotic chromosomes, DNA

DNA is made up of nucleosomes. Each 200 nucleotides in DNA are composed of a nucleosome. A nucleosome is a structure that consists of eight histones that are wrapped around by a DNA duplex. Extremely condensed portions of chromatin are called heterochromatin. A region of heterochromatin is never expressed since some of it remains condensed. Histones are proteins and carry a positive charge. They attract the phosphate groups (negatively charged) of DNA. In this way, histones direct the coiling of DNA.

5. Eight histone proteins have DNA coiled around it. This structure can be best described as a _____.

*   a. nucleosome
       b. centrosome
       c. nucleus
       d. gamete

6. The collection of chromosomes from a cell related to their shape, length and position of the centromere is referred to as which of the following?
       a. gamete
       b. nucleosome
*   c. karotype
       d. chromatin

A karotype is performed when conditions such as Down's syndrome are suspected. Down's syndrome is present when an extra chromosome is present on the 21st chromosome. This condition is also referred to as trisomy 21. Gametes are eggs or sperm.

7. Two copies of the same chromosome in the cells of the body are referred to as _____. Each chromosome contains _____ sister chromatids.
       a. homologous chromosomes, one
*   b. homologous chromosomes, two
       c. karotypes, two
       d. karotypes, one

8. The major growth phase in the cell cycle that takes the most time is called:
       a. S phase
*   b. G_1 phase
       c. G_2 phase
       d. mitosis

G_2 phase is the second growth phase where organelles in the cell replicate and get ready for chromosome separation. Condensation begins in the G_2 phase where the tight coiling of chromosomes occur. However, the G_1 and G_2 phases consist of no chromosomal replication. The G_2 phase takes about 6-8 hours for many eukaryotic organisms. The G_1 phase takes about 8-10 hours to occur. The S phase stands for the synthesis phase where chromosomes replicate. The S phase occurs in about 6-8 hours. Mitosis or the M phase is the phase in the cell cycle that consists of four stages, which are the following: prophase, metaphase, anaphase and telophase. Mitosis

only takes about an hour to occur. The G_1 phase, S phase, and G_2 phase make up interphase of the cell cycle.

9. Which portion of the cell cycle does not consist of interphase?
    a. S phase
*   b. mitosis
    c. G_1 phase
    d. G_2 phase

10. The first stage of mitosis is known as _____, and the last stage is referred to as _____.
    a. prophase, interphase
    b. interphase, anaphase
*   c. prophase, telophase
    d. metaphase, cytokinesis

Nuclear division that involves the increase in body cells occurs in mitosis. In anaphase, chromatin continues to condense and forms chromatids. Each individual chromatid is connected by a structure called a centromere. Centrioles are structures that help produce microtubules that form a spindle apparatus. Also, in anaphase, the centrioles and centromeres go to opposite poles of the cell. In metaphase, the centromeres go to the center of the cell. In anaphase, the centromeres break apart and the chromosomes split apart and go to opposite regions of the cell. In telophase, the mitotic spindle vanishes and the chromosomes revert back to chromatin form. Then cytokinesis occurs, which is when a cleavage furrow separates the parent cell (original cell) into two daughter cells.

11. Centromeres arrange at the center of the cell in _____ of mitosis. When two daughter cells form, _____ has begun.
*   a. metaphase, interphase
    b. prophase, telophase
    c. telophase, anaphase
    d. interphase, anaphase

12. The last stage of mitosis is referred to as _____. When a cleavage furrow forms and the cytoplasm separates into two new daughter cells, this is the process of _____.
    a. anaphase, cytokinesis

b. interphase, chromosome condensation
*   c. telophase, cytokinesis
d. prophase, chromosome condensation

The correct order of the cell cycle is in the following order: G_1-phase, S-phase, G_2-phase, prophase, metaphase, anaphase, telophase, and cytokinesis. Mitosis consists of prophase, metaphase, anaphase, and telophase.

13. The replication of chromosomes occurs in the _____ of interphase.
    a. G_1-phase
    b. G_2-phase
*   c. S-phase
    d. all of the above

14. Which of the following serves as a checkpoint for the continuation of the cell cycle?
    a. M checkpoint
    b. G_2 checkpoint
    c. G_1 checkpoint
*   d. all of the above

The G_1 checkpoint decides whether the cell cycle should go on to the S-phase. If G_1 checkpoint decides not to go to the S-Phase, the cell cycle may be put into the G_0 phase for a certain period of time. The G_0-phase is a resting phase. The G_2 checkpoint decides whether the cell cycle should go and begin mitosis. The M checkpoint is located at metaphase of mitosis. This checkpoint decides whether to end mitosis and begin cytokinesis. These checkpoints are useful because they slow down the cell cycle if a certain phase is taking longer than expected to be completed.

15. Which of the following specifically regulates the G_1 and G_2 checkpoints and is not a growth factor?
    a. epidermal growth factor
*   b. cyclins
    c. erythropoietin
    d. interleukin-2

16. Which of the following are types of enzymes that phosphorylate

microfilaments and histones so that the cell cycle can past checkpoint G_2 and enter mitosis?

*   a. cyclin-dependent protein kinases (Cdks)
         b. cyclins
         c. erythropoietin
         d. insulin-like growth factor

Cyclins help cyclin-dependent protein kinases (Cdks) to work as enyzmes. Cyclins are destroyed and created with each cell cycle. Erythropoietin is a growth factor that is required for red blood cells to grow into a mature state. Growth factors are proteins that aid in bringing on cell division.

17. _____ consists of G_2 cyclin that connects with cyclin-dependent protein kinase (Cdk). In this way, this type of growth factor will help the cell cycle start mitosis.

         a. Nucleosome
         b. Cytokinesis
         c. Erythropoietin
*   d. Mitosis-promoting factor (MPF)

18. Which is the shortest portion of the cell cycle?
         a. mitosis
         b. G_2-phase
         c. G_1-phase
*   d. cytokinesis

The entire cell cycle take approximately 22-24 hours to occur.

19. Which is a protein complex that is disk-like in shape, affixes to microtubules during mitosis and is connected to each centromere?
*   a. kinetochore
         b. karyotype
         c. chromatid
         d. nucleosome

20. Which phase of mitosis does a nuclear envelope reappear around sister chromatids?
         a. prophase
         b. metaphase
         c. anaphase

* d. telophase

Telophase is the last stage of mitosis where the spindle apparatus goes away. Right after telophase is cytokinesis and then the start of interphase.

21. The _____ phase is the non-growing stage of the cell cycle. If the cell cycle cannot go past the G_1 checkpoint, it may end up here.

    a. G_2

*   b. G_0

    c. M

    d. cytokinesis

22. Sex cells are referred to as _____. The fusion of gametes is labeled _____ and a new cell is produced. The cell that is made by this fusion is called a _____.

*   a. gametes, fertilization, zygote

    b. zygotes, fertilization, gamete

    c. chromosomes, meiosis, gamete

    d. gametes, meiosis, syngamy

Gametes in males are the sex cells that are called sperm. Gametes in females are the sex cells called eggs. Fertilization is also known as syngamy. There is only one set of chromosomes in the gametes of a male or female. This one set of chromosomes is known as haploid in number. However, other cells in the human body have two sets of chromosomes. Therefore, this makes them diploid.

23. Genes that bring on cell division are called _____ and have a role in developing cancer cells. Genes that stop cell division are called _____.

    a. Cdks, proto-oncogenes

    b. tumor-suppressor genes, Cdks

    c. tumor-suppressor genes, proto-oncogenes

*   d. proto-oncogenes, tumor-suppressor genes

24. Reproductive cell division occurs by the process of _____ in animals. This process creates _____ gametes.

    a. mitosis, haploid

*   b. meiosis, haploid

    c. meiosis, diploid

d. mitosis, diploid

Meiosis is characterized by a reduction division and equatorial division. Reduction division is called meiosis I, equatorial division is referred to as meiosis II. The zygote will continue to grow by the process of mitosis. The zygote is considered a diploid cell. This single zygote will form into an adult. The cells of the body are called somatic cells.

25. The pairing of homologous chromosomes is referred to as _____; this occurs in _____ of meiosis.

    a. crossing over, anaphase I

    b. synapsis, anaphase I

*   c. synapsis, prophase I

    d. crossing over, telophase I

26. Which stage of prophase I consists of the first formation of the synaptonemal complex?

    a. leptotene

*   b. zygotene

    c. pachytene

    d. diplotene

    e. diakinesis

Leptotene is the first stage of prophase I. In this stage, chromosomes condense or are compacted firmly. Zygotene, the second stage of prophase I, is where the synaptonemal complex first forms. The synaptonemal complex is a structure that makes crossing over between chromosomes much easier. Pachytene is stage three of prophase I. Actual crossing over of chromosomes occurs in this stage. The crossing over makes a structure called a chiasma. Diplotene is stage four of prophase I. This is where the synaptonemal complex breaks apart. Also, cell growth advances thoroughly in this stage. Diakinesis is the fifth stage of prophase I. In this stage, the chiasmata go to the terminal part of the chromosomes. Later, the terminal chiasmata will then bind the homologous chromosomes jointly in metaphase I.

27. Crossing over between chromosomes occurs in which stage of prophase I of meiosis?

    a. leptotene

    b. diplotene

    c. diakinesis

*   d. pachytene

28.   Homologous chromosomes are aligned at the metaphase plate in metaphase I.   Which is not a true statement regarding metaphase I in meiosis?
      a. Spindle microtubules can bind to kinetochore proteins on the outside of each centromere.
*   b. Spindle microtubules can bind to kinetochore proteins on the inside of the centromeres.
      c. Homologous chromosomes are bound together at their terminal portions by terminal chiasmata.
      d. Chromatids are held together by chiasmata.
Spindle microtubules can bind to kinetochore proteins just on the outside of each centromere in metaphase I of meiosis. However, kinetochore proteins can bind to the inside or outside of each centromere in metaphase of mitosis.

29.   Crossing over between homologous chromosomes occurs in which stage of prophase I?
*   a. pachytene
      b. leptotene
      c. diplotene
      d. diakinesis

30.   Which of the following occurs in metaphase I?
*   a. Pairs of homologues line up at the metaphase plate.
      b. Crossing over occurs forming a structure called the chiasma.
      c. The synaptonemal complex forms.
      d. Each homologue pair drifts apart and each homologue in each pair goes to one pole of the cell.
Crossing over forms a structure called the chiasma and happens in the stage of pachytene of prophase I. The synaptonemal complex first forms in the zygotene stage of prophase I. The synaptonemal complex existing in the pachytene stage falls apart in the diplotene stage of prophase I. Each pair of homologue drifts apart and each homologue in each pair goes to one pole of the cell in anaphase I of meiosis.

31.   All of the following are true statements regarding telephone I in meiosis except:

a. Nuclear membrane reappears surrounding the nucleus of each of the two daughter cells.

b. After telephone I, cytokines is may occur.

c. Anaphase I occurs before telephone I.

* d. Crossing over occurs in telophase I.

32. Which is true regarding the interphase that occurs right after meiosis I?
* a. Replication of chromosomes does not occur in the interphase that occurs right after meiosis I.

b. Homologous chromosomes line up at the metaphase plate during interphase of mitosis.

c. Mitotic spindle forms in interphase of mitosis but not in interphase of meiosis.

d. The synaptonemal complex breaks apart in interphase of meiosis but not in interphase of mitosis.

Replication of chromosomes occurs in the S-phase of interphase in mitosis. However, there is a replication of chromosomes in the interphase before prophase I of meiosis.

33. Reduction division occurs in _____, and equatorial division occurs in _____. The replication of chromosomes does not occur in between these two divisions.

a. meiosis II, meiosis I
* b. meiosis I, meiosis II

c. mitosis, meiosis II

d. meiosis I, meiosis I

34. Meiosis II is most similar to _____. There are _____ haploid cells that develop from meiosis II.

a. meiosis I, four
* b. mitosis, four

c. mitosis, two

d. meiosis I, two

Meiosis II produces four haploid cells in both males and females. In human males, these four haploid cells grow into four sperm cells. In human females, only one of the four haploid cells will grow into an egg or ovum. The other haploid cells will just grow into polar bodies, which do not become eggs.

35. Spindle fibers first connect to both sides of the centromere in _____.
      a. prophase II
      b. telophase II
*   c. metaphase II
      d. anaphase II

36. The centromere first disconnects in _____ of meiosis II. Haploid sets of chromosomes in four cells occur in _____ of meiosis II.
*   a. anaphase II, telophase II
      b. metaphase II, anaphase II
      c. prophase II, metaphase II
      d. prophase II, telophase II
Each nuclear envelope breaks apart and new spindle starts to be made in anaphase II. The spindle fibers connect to both sides of the centromere in metaphase II. The centromeres disconnect and the sister chromatids transfer to opposite poles in anaphase II. The nuclear envelope is made, and it surrounds the four sets of daughter chromosomes in meiosis II.

37. Which of the following is a false statement regarding asexual reproduction?
      a. The offspring will have the same genetic information as the parent.
      b. Binary fission is an example of asexual reproduction.
      c. Some animals can reproduce asexually.
*   d. The cell produced from asexual reproduction is not identical to the parent.

38. The growth into an adult organism from an unfertilized egg is called:
*   a. parthenogenesis
      b. crossing over
      c. meiosis I
      d. meiosis II
Arthropods and some fish and lizards use parthenogenesis to reproduce. Parthenogenesis is a form of reproduction without sex.

39. In Mendel's cross-fertilization experiments, the white pea flower that resulted from crossing of a white pea flower and a purple pea flower in the

second generation of offspring was labeled a _____ trait. The first generation of offspring is known as _____.

      a. dominant, first filial (F_1 generation)

      b. recessive, dominant generation

      c. dominant, dominant generation

*   d. recessive, first filial (F_1 generation)

40. The first generation of offspring from the Mendel experiments regarding the cross-fertilization of the white pea flower and the purple pea flower resulted in which of the following?

*   a. purple pea flowers only

      b. white pea flowers only

      c. approximately 76% purple pea flowers and 24% white pea flowers

      d. approximately 24% purple pea flowers and 76% white pea flowers

In the second generation of offspring approximately 76% were purple pea flowers and 24% were white pea flowers. This is a ratio of approximately 3:1. The dominant trait was the purple pea flowers, which occurs at a rate of 76% in the second generation; the recessive trait was the white pea flowers, which occurs at a rate of 24% in the second generation.

41. Mendel's phenotype ratio can best be depicted as _____ in the F_2 generation of purple pea flowers, and his genotype ratio can be best depicted as _____.

*   a. 3:1, 1:2:1

      b. 1:2:1, 3:1

      c. 4:1, 1:2:1

      d. 1:2:1, 4:1

42. The precise position of a gene on a chromosome is referred to as which of the following?

      a. genotype

      b. phenotype

*   c. locus

      d. recessive trait

Genotype is the complete genetic makeup of a person. It is the entire amount of alleles in an individual. The phenotype is the actual physical expression of traits in an organism.

43. An organism that is diploid and has two different alleles of a gene on a pair of homologous chromosomes is referred to as a _____; an organism that is diploid that has the same alleles of a gene on a pair of homologous chromosomes is referred to as a _____.

    a. heterozygote, heterozygote
*   b. heterozygote, homozygote
    c. homozygote, homozygote
    d. homozygote, heterozygote

44. Mendel's experiment regarding the cross-fertilization of the white pea flower with the purple pea flower resulted in which of the following in the F_2 generation?

    a. 25% are *PP*
    b. 50% are *Pp*
    c. 25% are *pp*
*   d. all of the above

PP represents the dominant homozygotes. *Pp represents the heterozygotes, and pp represents the recessive homozygotes.* The large *P* refers to the dominant allele, and the *p* refers to the recessive allele. These results are achieved by making a Punnett square using *PP* as the purple pea flower and *Pp* as the white pea flower for the F_1 generation. Combining these two plants with *PP* and *Pp* will give 100% with the *Pp* configuration for the generation one or F_1 generation. Referring back to this question, the F_2 generation can be attained by using the *Pp* configuration from F_1 generation result. Using two purple pea flowers with *Pp* configuration (from the F_1 generation result) in a new Punnett square will result in 25% *PP*, 50% *Pp*, and 25% *pp*. This is the genotype ratio of 1:2:1. However, three purple pea flowers and one white pea flower result in F_2 generation. This is the phenotype ratio of 3:1.

45. Individual alleles that have two or more effects on a phenotype are labeled:

    a. epistasis
*   b. pleiotrophic
    c. syntenic
    d. centimorgan

46. Which of the following is the primary reason Mendel used a testcross?

*   a. to find out the genotype of a dominant phenotype
     b. to find out the phenotype of a dominant genotype
     c. to find out how many chromosomes are in purple pea plants
     d. to test if a Punnett square really works

A testcross is performed by crossing a dominant phenotype (purple pea flower) with a homozygous controlled white pea flower. This is performed to find out the unknown genotype of the dominant phenotype. In other words, to find out if the dominant phenotype has a genotype that is homozygous (*PP)* or heterozygous *(Pp)*. Testcrosses can be performed in other living things as well. The Punnett square was not created by Mendel. It was created by Reginald Crundall Punnett. A testcross does not have a main purpose of proving the validity of a Punnett square.

47.   Mendel's testcross of a heterozygous purple pea plant *(Pp)* with homozygous recessive white pea plant (*pp*) proved which of the following?
*   a. Recessive to dominant traits occur in a 1:1 ratio.
     b. Recessive to dominant traits occur in a 2:1 ratio.
     c. Dominant phenotypes are always homozygous.
     d. Dominant phenotypes are always heterozygous.

48.   Mendel's dihybrid cross experiments with round, yellow seeds and wrinkled, green seed demonstrate a ratio of which of the following?
*   a. 9:3:3:1
     b. 1:1
     c. 1:3
     d. 4:2

Since round, yellow seeds are two traits, it is represented as *RRYY*. Since wrinkled, green seeds are two traits, it is represented as *rryy*. *RRYY* is first crossed with *rryy* to make the F_1 generation that has the genotype of *RrYy* or the phenotype of all round, yellow seeds. These round, yellow seeds that resulted with the new genotype of *RrYy* are considered dihybrids. These dihybrids self-fertilize and create the following: 9 round, yellow peas, 3 round, green peas, 3 wrinkled, yellow peas and 1 wrinkled, green pea. This gives the ratio of 9:3:3:1.

49. When one gene obstructs another gene's expression, this occurrence is called _____.
     a. genotype

*   b. epistasis
     c. locus
     d. allele

50. Male sex chromosomes are identified as _____ and female sex chromosomes are identified as _____.
     a. XX, YY
     b. XY, XY
*   c. XY, XX
     d. XX, XY

There are 23 pairs of chromosomes in the human body. The 23rd pair is called sex chromosomes. The other 22 pairs are called autosomes. Also, the male fruit fly (Drosophilia melanogaster) has sex chromosomes represented as XY. The female fruit fly has sex chromosomes represented as XX.

51. The inactivated X chromosome in a female embryo is better known as a(n) _____.
     a. monosomic
     b. allele
     c. autosome
*   d. Barr body

52. Humans that have lost one chromosome in a pair of autosomes are referred to as:
*   a. monosomics
     b. trisomics
     c. Barr bodies
     d. alleles

Humans that have three autosomes instead of two in a particular pair are called trisomics. For example, Trisomy 21 is also known as Down's Syndrome. This is caused by an extra autosome on the 21st pair of autosomes. Down's syndrome causes mental retardation, low set ears, flat face and short fingers to name a few of its symptoms. Down's syndrome has been recorded as occurring in about 1 out of 725 births.

53. Which of the following is a type of genetic recombination?
     a. locus

b. Barr body
*　　c. crossing over
　　　d. allele

54. An individual with Klinefelter syndrome is _____ and has a sex chromosome configuration of _____.
　　　a. female, XXX
　　　b. female, XXY
*　　c. male, XXY
　　　d. male, YYX
A male with Klinefelter syndrome will be unable to have children. Also, this male will be feminine in appearance. An individual with a sex chromosome configuration of XXX will be female but some will be unable to reproduce and have children. However, women with XXX configuration can have normal intelligence levels. XXX and XXY are considered nondisjunctions of sex chromosomes. Nondisjunction pertains to a mistake occurring in meiosis. Nondisjunction can cause added or deleted sex chromosomes.

55. Which of the following best identifies a sterile female with a webbed neck and a sex chromosome configuration of XO?
　　　a. Klinefelter syndrome
*　　b. Turner syndrome
　　　c. Down's syndrome
　　　d. Tay-Sachs disease

56. Which of the following is a recessive genetic disorder that causes excessive mucus in the organs of the body?
　　　a. sickle cell anemia
　　　b. Tay-Sachs disease
　　　c. muscular dystrophy
*　　d. cystic fibrosis
Cystic fibrosis is a disease of excessive mucus accumulation in the lungs, pancreas and heart that usually causes death before middle age. Sickle cell anemia is a recessive genetic disorder usually in African-Americans where the blood cell is actually irregular in shape. The normal red blood cell is circular in shape. In sickle cell anemia, these irregular red blood cells or sickle cells have an abnormal form of hemoglobin. Normal hemoglobin is

needed because it functions to bring oxygen through out the body. Tay-Sachs disease is a recessive genetic disorder that occurs in Eastern and Central European people of Jewish descent. In this disorder, gangliosides build up excessively in the central nervous system usually leading to death by age of five. Gangliosides are the products of fat metabolism.

57. Which of the following is a dominant genetic disorder that causes brain destruction that usually begins after the age of 30?

      a. cystic fibrous

*   b. Huntington's disease

      c. hemophilia

      d. sickle cell anemia

58. When genes are very approximate to one another on a chromosome, they will not assort separately. These genes are referred to as being:

*   a. linked

      b. Barr bodies

      c. autosomes

      d. monosomics

A three-point cross includes three genes that are linked. Remember that genetic recombination between genes occurs greater the farther away the genes are from one another.

59. _____ is used to measure the distance between genes, and one map unit is called a _____.

      a. Centimorgan, crossing over

      b. Three-point cross, genetic map

*   c. Genetic map, centimorgan

      d. Autosome, three-point cross

60. Which of the following is a sex-linked recessive condition that causes an inability for the blood to clot?

      a. cystic fibrosis

*   b. hemophilia

      c. muscular dystrophy

      d. Tay-Sachs disease

There are mutations in certain proteins in the blood that cause this condition. Genes that are responsible for blood clotting are mostly on

autosomes. However, there are two genes on the X chromosome that are responsible for blood-clotting also. These two genes are identified as VIII and IX. If either VIII or IX are mutants, hemophilia will result. Royal families have a history of this condition. Hemophilia VIII is identified as hemophilia A, and hemophilia IX is identified as hemophilia B. Males usually inherit this condition from their mother.

Genetics

1. According to the biologist Joachim Hammerling, the nucleus is located in the _____ of the green algae Acetabularia.
     a. stalk
     b. cap
*   c. foot
     d. leaf

2. Which of the following uses RNA as its genetic material?
     a. humans
     b. bacteria
*   c. viruses
     d. all of the above
Humans, bacteria and viruses use DNA as its genetic material. However, there are some viruses that can use RNA as its genetic material.

3. Which of the following is most consistent with Chargaff's rules regarding DNA?
*   a. The quantity of adenine should equal the quantity of thymine; the quantity of cytosine should equal the quantity of guanine.
     b. Phosphodiester bonds are absent in DNA.
     c. Purines are adenine and guanine; pyrimidines are thymine and cystosine.
     d. Purines and pyrimidines are nitrogen-containing bases

4. In DNA and RNA, which of the following is part of a nucleotide?
     a. phosphate group
     b. nitrogen-containing base
     c. sugar
*   d. all of the above

DNA is made up of a five carbon sugar called deoxyribose. In addition, it is made up of a phosphate group and a nitrogen-containing base. RNA is made up of a five carbon sugar called ribose, a phosphate group and a nitrogen-containing base. However, in RNA, the base, thymine, is substituted by the base uracil. DNA can have the following nitrogen-containing bases: adenine (A), guanine (G), thymine (T), cytosine (C). RNA can have the following nitrogen-containing bases: adenine (A), guanine (G), uracil (U) and cytosine (C).

5. All of the following are possible nitrogen-containing bases in RNA except:

     a. cytosine

     b. adenine

 *   c. thymine

     d. uracil

6. Each DNA nucleotide is connected to another DNA nucleotide by a _____ phosphate group linking to a _____ hydroxyl group.

 *   a. 5′, 3′

     b. 2′, 3′

     c. 3′, 1′

     d. 4′, 5′

The sugar molecule, deoxyribose, consists of five carbons. Each carbon is labeled 1, 2, 3, 4, 5 in a clockwise path. The prime symbol (′) in 5′, for example, means that carbon is found in a sugar molecule. In DNA or RNA, 5′ phosphate group means that a phosphate group occurs on the fifth carbon molecule of deoxyribose. In addition, 3′ hydroxyl group means that a hydroxyl group occurs on the third carbon molecule of deoxyribose.

7. In DNA or RNA, dehydration synthesis is a reaction that occurs between a phosphate group of one nucleotide trying to connect with the hydroxyl group of another nucleotide. The resulting connection is referred to as a _____. Hence, this connects nucleotides together.

     a. hydrogen bond

     b. non-covalent bond

 *   c. phosphodiester bond

     d. triple bond

8.   Which of the following best identifies the chemical structure of a phosphodiester bond in DNA or RNA?

    a. O—H—O

    b. C—H

    c. H—O

*   d. P—O—C

This phosphodiester bond is directly connected to the sugar, deoxyribose, in DNA, and ribose in RNA.

9.   In RNA, adenine is linked to _____, and in DNA, guanine is linked to _____.

*   a. uracil, cytosine

    b. cytosine, thymine

    c. thymine, cytosine

    d. guanine, adenine

10.   Which of the following is not true regarding the structure of DNA?

    a. It's a double helix.

    b. On end of DNA has a structure of 5′ to 3′ and the other end has a structure of 3′ to 5′.

    c. Guanine is attached to cytosine and adenine is attached to thymine.

*   d. There are two hydrogen bonds between cytosine and guanine; there are three hydrogen bonds between adenine and thymine.

There are actually three hydrogen bonds between cytosine and guanine, and two hydrogen bonds between adenine and thymine. Watson and Click used the results of Franklin's X-ray experiments on the structure of DNA and Chargaff's rules to physically show the DNA structure.

11.   If one single strand of DNA is in a sequence of 5′-CATGTTA-3′, what would be its complementary strand identified as?

    a. 3′-ATGTTAT-5′

*   b. 3′-GTACAAT-5′

    c. 5′-GTAUAAT-3′

    d. 5′-ATGTTAT-3′

12.   Which of the following is not a true statement regarding DNA?

    a. DNA replication is semiconservative.

    b. The two strands of DNA in a double helix are mirror images of

each other.

    c. Hydrogen bonds connect each strand of DNA together to form a double helix; these bonds are found between A and T and G and C.

*   d. If DNA strands are mirror images of each other, the DNA molecule will always break apart.

In DNA, there are hydrogen bonds between T and A and G and C, which are base pairs. Each T connects with A. In addition, G connects with C in DNA.

13. There are _____ hydrogen bonds between T and A in DNA. There are _____ hydrogen bonds between G and C in DNA.

*   a. two, three
    b. three, two
    c. one, four
    d. four, one

14. Which is the process where DNA encoded sequences are copied to a strand of RNA?

    a. translation
    b. base-pairing
*   c. transcription
    d. cell death

Transcription occurs by DNA encoded sequences being copied to messenger RNA or mRNA. RNA polymerase is an enzyme that catalyzes this transcription. Translation is the way where base sequences in mRNA determine the amino acid sequence of a protein.

15. Which of the following has the most primary role in DNA replication?

*   a. DNA polymerase III
    b. DNA polymerase I
    c. primase
    d. helicase

16. Which of the following unwinds the double helix structure of DNA?

    a. primase
*   b. helicase
    c. DNA polymerase III
    d. DNA ligase

Primase makes RNA primers. Primase is a type of RNA polymerase. DNA polymerase III replicates DNA. DNA ligase connects each end portion of DNA of the lagging strand.

17. In DNA, which of the following extends toward the replication fork and is being assembled by adding nucleotides to its 3′ end?
      a. lagging strand
      b. DNA ligase
*   c. leading strand
      d. helicase

18. _____ synthesis occurs on the leading strand of DNA. _____ synthesis occurs on the lagging strand of DNA. However, the whole process is referred to as semidiscontinuous replication.
      a. Discontinuous, Continuous
      b. Discontinuous, Discontinuous
*   c. Continuous, Discontinuous
      d. Continuous, Continuous

In continuous synthesis, first primase makes an RNA primer. The RNA primer is approximately 10 RNA nucleotides on the leading strand. Then DNA polymerase III inserts nucleotides to the 3′ end of the leading strand. Finally, DNA polymerase I takes away the RNA primer and puts DNA nucleotides down to help make the new DNA strand. In discontinuous synthesis, primase makes RNA primer, which goes on the 5′ end of the lagging strand. However, the lagging strand is made discontinuously. Since it is made discontinuously, fragments are formed called Okazaki fragments. Okazaki fragments are made of nucleotides. Polymerase III makes these Okazaki fragments. Now it is up to DNA ligase to connect each of these Okazaki fragments to the lagging strand. Also, DNA polymerase I takes away the RNA primer and puts DNA nucleotides down to help make the new DNA strand.

19. Which of the following is found on a lagging strand of DNA but not on a leading strand of DNA?
*   a. Okazaki fragments
      b. DNA polymerase III
      c. DNA polymerase I
      d. RNA primer

20. Which of the following is an enzyme that alleviates torque during DNA replication?

      a. DNA polymerase III

*   b. DNA gyrase

      c. helicase

      d. RNA primer

Helicase brings apart or unwinds the double helix of DNA. Torque is the twisting motion that occurs in a DNA strand during replication.

21. Which of the following is a set of three consecutive nucleotides and is a triplet code?

*   a. codon

      b. DNA

      c. RNA

      d. tRNA

22. Which of the following determines which amino acids are made into polypeptides in the ribosome?

      a. transfer RNA

      b. RNA polymerase I

      c. DNA ligase

*   d. messenger RNA

During transcription, RNA polymerase makes mRNA sequences complementary to a certain DNA molecule. In translation, mRNA sequences determine the amino acid sequence of a protein. Transcription occurs before translation.

23. In the cytoplasm of humans, where does translation first occur?

      a. anticodon

*   b. start codon

      c. end codon

      d. all of the above

24. Which of the following is transcribed on DNA in transcription?

*   a. template strand

      b. coding strand

      c. promoter sites

      d. triplet code

The coding strand is the strand of DNA that is not transcribed. Promoter sites are referred to as RNA polymerase binding sites. These promoter sites are found on the template strand of DNA. The coding strand is called the sense strand; the template strand is called the antisense strand.

25. Gene transcription begins when which of the following occurs?
    a. Molecule of mRNA binds to rRNA in a ribosome.
    b. Molecule of rRNA binds to tRNA in a ribosome.
*   c. RNA polymerase binds to the promoter sites on the template strand of DNA.
    d. RNA polymerase binds to the promoter sites on the coding strand of DNA.

26. The -35 sequence is a six-base sequence that is found in bacterial promoter sites. Which is another common six-base sequence that is made up of bacterial promoter sites?
    a. 0 sequence
*   b. -10 sequence
    c. +35 sequence
    d. -954,125 sequence
-35 sequence means that this promoter site is 35 nucleotides away from the actual place where transcription first begins. The -35 sequence has the six-base sequence of TTGACA. The -10 sequence has the six-base sequence of TATAAT.

27. Which of the following is a false statement regarding RNA transcription?
    a. RNA transcription begins with ATP.
    b. The transcription bubble contains DNA with RNA polymerase and also RNA transcript.
*   c. Transcription is the formation from RNA to protein.
    d. RNA polymerase has a major influence on RNA transcription.

28. Which of the following stops gene transcription?
*   a. GC hairpin
    b. transcription bubble
    c. RNA polymerase
    d. mRNA

At the last part of a gene, there are stop sequences that contribute to the following occurrences: breaking apart of the RNA-DNA hybrid in the transcription bubble, liberation of DNA by RNA polymerase and the unwinding of the transcription bubble. The GC hairpin is made of G-C base pairs that causes gene transcription to cease.

29. Which of the following is a complementary three-nucleotide sequence found in tRNA?
     a. DNA ligase
     b. helicase
*   c. anticodon
     d. GC hairpin

30. Which of the following directly helps tRNA connect with certain amino acids?
     a. template strand
*   b. aminoacyl-tRNA
     c. coding strand
     d. none of the above

The template strand is one of two strands found on DNA that is used for RNA transcription. The coding strand is the other strand found on DNA that is not used for RNA transcription. Aminoacyl-tRNA is an activating enzyme. In RNA translation, activating enzymes must match to a certain anticodon on tRNA and a certain amino acid.

31. Which of the following understands the mRNA codon message?
     a. nucleus
     b. DNA
     c. coding strand
*   d. activating enzymes

32. In genes, there are areas called _____, which do not code for synthesis of a section of a protein. In addition, regions in DNA that do code for parts of a polypeptide are called _____.
*   a. introns, exons
     b. exons, introns
     c. codons, introns
     d. exons, anticodons

RNA transcript in the beginning contains introns and exons. In RNA splicing, introns are cut out of the primary transcript. In fact, about 75% of the original RNA are cut out in RNA splicing. The 25% of the original RNA that is left over is spliced or joined together. Now mRNA is made. This final mRNA will then enter translation.

33. Which is the first step in protein synthesis?
     a. RNA splicing first begins.
*   b. RNA polymerase transcribes RNA out of DNA.
     c. The synthesis of mRNA begins.
     d. mRNA goes to the cytoplasm from the nucleus.

34. In protein synthesis, mRNA will go to the cytoplasm from the nucleus. Then mRNA in the cytoplasm will become connected with ribosomal subunits. In the next step, activating enzymes will aid _____ to become connected with certain amino acids.
     a. RNA polymerase
     b. DNA
     c. nuclear membrane
*   d. tRNA

Protein synthesis first begins in the nucleus where RNA polymerase transcribes RNA out of DNA. Step two consists of the making of mRNA. This is done through RNA splicing. Step three consists of mRNA going to the cytoplasm from the nucleus and then connecting with ribosome subunits. Step four includes activating enzymes helping tRNA get connected with certain amino acids. In step five, tRNA and its connected amino acids first go on a portion of the ribosome referred to as the A site. tRNA will then exit the ribosome by way of the portion of the ribosome called the E site. The remaining sixth step will include the growth of this polypeptide chain until the finishing point of this protein.

35. Primarily, transcriptional control of certain genes is regulated in eukaryotes and bacteria by which of the following?
     a. nucleus
*   b. RNA polymerase
     c. cytoplasm
     d. tRNA

36. Regulatory proteins also can control gene expression in DNA. This is performed by _____ inserting into the _____ of DNA.
*   a. DNA-binding motifs, major groove
      b. tRNA, DNA-binding motifs
      c. mRNA, DNA-binding motifs
      d. DNA-binding motifs, RNA polymerase

There is a minor groove and a major groove in the helical structure of DNA. The major groove is the location where regulatory proteins can connect. In this way gene expression can be regulated also. Motifs are structures. So each DNA-binding motif has a different structure. These DNA-binding motifs consist of regulatory proteins that control gene expression in DNA.

37. Which is the most common type of DNA-binding motif?
*   a. helix-turn-helix motif
      b. zinc finger motif
      c. homeodomain motif
      d. leucine zipper motif

38. Which portion of the helix-turn-helix is found in the major groove of DNA?
      a. zinc fingers
      b. leucine zipper
*   c. recognition helix
      d. tRNA

The helix-turn-helix consists of a recognition helix (helical structure) and the other helix (helical structure) is on the outside of the DNA structure. Helix-turn-helix motif are two α–helical segments. This helix-turn-helix is composed of regulatory proteins. The leucine zipper is another DNA-binding motif. It has a Y structure, and it fits into the major groove of DNA. Zinc finger is another kind of DNA-binding motif. This type of motif utilizes zinc atoms to help bind α–helical segments to the major groove in DNA.

39. Which of the following DNA-binding motifs is Y-shaped and does not use zinc atoms?
*   a. leucine zipper
      b. helix-turn-helix
      c. zinc finger

d. homeodomain

40. Bacteria stop transcription of genes from occurring by using _____, which are proteins that connect to the promoter site. In this way, RNA polymerase will not be able to bind with the promoter site.
    a. promoters
    b. activators
*   c. repressors
    d. all of the above
Bacteria have activators, which are regulatory proteins that help to initiate transcription. This is done by the activator enhancing the ability of the promoter to connect or bind with RNA polymerase. The actual binding location of RNA polymerase is referred to as a promoter.

41. Which is a group of functionally related genes that are transcribed into a mRNA molecule?
    a. codon
    b. intron
*   c. operon
    d. repressor

42. _____ is a regulatory protein that binds to a structure found on DNA called an _____.
    a. Enhancer, activator
*   b. Activator, enhancer
    c. Operator, activator
    d. Codon, enhancer
In eukaryotes and few bacteria, there are distant locations referred to as enhancers where regulatory proteins bind to DNA. Repressors bind to a location on DNA called the silencer. In this way, repressors stop the binding of enhancers with activators. In addition, this stops or decreases the rate of RNA transcription.

43. Which of the following is a regulatory protein that connects with the silencer on DNA and can prevent RNA transcription from occurring?
    a. intron
    b. enhancer
    c. activator

*   d. repressor

44. Which of the following is a transcription factor that brings signals from the activator to basal factors located on the RNA polymerase?
*   a. coactivator
        b. enhancer
        c. silencer
        d. repressor
The coactivator is connected to the activator. The basal factors help RNA polymerase to bind with DNA. This is performed by the basal factors aiding the RNA polymerase to the proper position on DNA. Then when the RNA polymerase is in the proper position, the basal factors detach from it. Then RNA polymerase can carry out the transcription of mRNA.

45. Coactivators and _____ are considered transcription factors. They help in the binding of RNA polymerase with DNA.
        a. repressors
        b. enhancers
*   c. basal factors
        d. silencers

46. Coactivators bind to activators and _____, which is connected to the _____.
        a. repressor, enhancer
        b. repressor, TATA-binding protein
        c. enhancer, basal factors
*   d. TATA-binding protein, basal factors
In addition, the basal factors are connected to the RNA polymerase. Basal factors are no longer connected with RNA polymerase when it is let go to undergo the transcription of mRNA.

47. _____ is a regulatory protein that activates RNA polymerase by touching it. This is accomplished by DNA specific looping, which makes this possible.
        a. intron
*   b. enhancer
        c. repressor
        d. silencer

48. Before translation is preformed, introns are taken out from the primary RNA transcript in the process referred to as:
*   a. RNA splicing
        b. transcription
        c. gene expression
        d. translation repression

Approximately 91% of the RNA transcript are made of introns. RNA splicing is also known as RNA processing. The introns are cut apart and put together by a structure referred to as the spliceosome. Translation repression is a process where translation repressor proteins prevent translation by connecting to the starting point of the RNA transcript. In this way, the RNA transcript will not able to bind with a ribosome.

49. Which of the following is a type of posttranscription control of gene expression?
        a. selectively degrading mRNA transcripts
        b. RNA splicing
        c. translation repression
*   d. all of the above

50. Mutations in _____ will be passed to future generations.
*   a. germ-line tissue
        b. somatic tissue
        c. skin tissue
        d. nerve tissue

Germ-line tissues are the gametes. Somatic cells are the various tissues of the body. However, somatic cell mutations cannot be passed down to future generations. Mutations are any alteration of genes.

51. Changes in one or two base pairs of a genetic message are labeled which of the following?
        a. translation regression
        b. metastasis
*   c. point mutation
        d. carcinogens

52. Which of the following is an example of a mutagen?
        a. ultraviolet radiation

b. ionizing radiation

c. chemicals

* d. all of the above

DNA is injured by mutagens. Ultraviolet radiation causes the so called pyrimidine dimer. This results when two pyrimidines on a DNA strand get bonded together. This unnatural bond can cause a bend in the double helix of DNA. Thus, replication of the double helix of DNA will be stopped. Ionizing radiation will damage the phosphodiester bonds of DNA. This damage is also known as a double-strand break. This break is accomplished by ionizing radiation going into the cells of the individual and causes free radicals to be released that damage DNA. One form of ionizing radiation comes in the form of x-rays. Chemicals can cause mispairing of bases in DNA.

53. Pyrimidine dimers are usually caused by which type of DNA damage?

a. transposition

b. chemicals

c. ionizing radiation

* d. ultraviolet radiation

54. Which is a section of mutation defined as the movement of genes at one genome to another location?

* a. transposition

b. inversion

c. spontaneous error

d. all of the above

Inversion describes a point of reference for a section of a chromosome that gets turned around and is backwards. Inversions most of the time do not change the gene expression. Spontaneous errors in mutations occur very infrequently. An example of a spontaneous error in mutations is slipped mispairings. When homologues line up and there is a mispairing of DNA sequences, one of the pairs will form a loop. This portion of loop will be cut out in order to get rid of the faulty mispairing nucleotide sequence. Losing this loop is called a deletion. This is a deletion of nucleotides. These deletions will then cause frame-shift mutations. In other words, now some of the bases will be deleted leading to mistakes in the genetic message.

55. Which of the following has a primary role in insertional inactivation, a type of mutation?

      a. double-stand break

*   b. transposon

      c. pyrimidine dimer

      d. slipped mispairing

56. Placement of a transposon into a gene is a type of _____, which occurs in _____, a type of mutation in DNA.

*   a. transposition, insertional inactivation

      b. transposition, deletion

      c. homologue pairing, insertional inactivation

      d. homologue pairing, deletion

Transposon is a moveable piece of DNA that goes from one genome and then sticks into another. When a transposon sticks into another gene, it makes it (the gene) inactive; this is referred to as insertional inactivation.

57. Cancerous tumors that are derived from _____ tissue are referred to as sarcomas.

      a. muscle

      b. bone

      c. connective

*   d. all of the above

58. Which is a type of cancer that derives from epithelial tissue?

*   a. carcinoma

      b. sarcoma

      c. carcinogen

      d. benign tumor

Carcinogens are types of radiation and certain chemicals that cause cancer. A benign tumor is not cancerous. Sarcomas are cancerous and occur in the bone (i.e., osteosarcoma), connective tissue and in the muscle.

59. Which is not a true statement regarding cancer?

      a. Oncogenes are the types of genes that cause cancer.

      b. Cancer is uncontrollable cell production.

      c. The spread of cancer to many parts of the body is referred to as metastasis.

*   d. Proto-oncogenes never become oncogenes.

60.   Which of the following is a type of oncogene?
      a. erb-B
      b. RET
      c. N-ras
*   d. all of the above
N-ras is responsible for leukemia, RET causes cancer of the thyroid and, erb-B causes breast cancer.

61.   _____ and _____ are two ways genetic recombination can occur.
*   a. Gene transfer, reciprocal recombination
      b. Slipped mispairing, satellite DNA
      c. Cell proliferation, satellite DNA
      d. Oncogenes, introns

62. All of the following are types of reciprocal recombinations except:
      a. gene conversion
      b. unequal crossing over
      c. crossing over
*   d. conjugation
Conjugation and transposition are types of gene transfers. Gene conversion includes the pairing of homologous chromosomes and one is made to look exactly like the other. Unequal crossing over shows the way to gene replication and deletion of genes. Crossing over occurs during meiosis in eukaryotes. Reciprocal recombination occurs in eukaryotes and gene transfers occur in prokaryotes and eukaryotes.

63. Which of the following is an extrachromosomal DNA fragment that is circular in appearance and is important in gene transfer?
      a. transposon
      b. oncogene
*   c. plasmid
      d. mutagen

64. Plasmids and _____ move genes to new sections on chromosomes. In this way, gene transfer is carried out.

*   a. transposons
       b. introns
       c. exons
       d. satellite DNA

Plasmids are created through reciprocal exchange. Plasmids are involved with gene transfer by the process of conjugation. Conjugation is the physical exchange of genes between bacteria.

65. Which of the following in an enzyme that aids in the insertion of transposon into a gene?
       a. protein kinase
*   b. transposase
       c. RNA polymerase
       d. DNA polymerase

66. Gene mobilization and _____ are two aspects of gene transfer by transposition.
       a. conjugation
*   b. insertional inactivation
       c. pilus
       d. rolling-circle replication

In a bacterial cell, a pilus forms a conjugation bridge between another bacterial cell. The bacterial cell that forms a conjugation bridge has a plasmid. The plasmid is replicated by a process referred to as rolling-circle replication and then transferred through the conjugation bridge to the other bacterial cell. In this way, this other bacterial cell now has a complete copy of the plasmid. Conjugation is another method of gene transfer.

67. Which are short sections of nucleotides that are duplicated millions of times and are found at the ends of chromosomes?
       a. transposons
       b. multigene families
       c. tandem clusters
*   d. satellite DNA

68. Which of the following consists of genes that are almost the same in nucleotide sequence, are active and occur in large groups of hundreds?
*   a. tandem clusters

b. satellite DNA

c. single-copy genes

d. pseudogenes

Single-copy genes will be later reproduced again in the cell cycle. Genes that are inactive by cause of mutation are labeled pseudogenes.

69. Which of the following can specifically result when trinucleotides become repeated too many times leading to disorders such as fragile X syndrome?

    a. conjugation

*   b. trinucleotide repeats

    c. rolling-replication

    d. RNA polymerase

70. Which of the following is a stage of an experiment regarding genetic engineering?

    a. production of recombinant DNA

    b. cloning

    c. DNA cleavage

    d. screening

*   e. all of the above

The stages of an experiment regarding genetic engineering occur in the following order: DNA cleavage, production of recombinant DNA, cloning and screening. The first stage involves utilizing a restriction endonuclease to cleave or cut DNA into smaller sections. In the second stage, the resulting sections of DNA are put into the plasmids. This second stage results in the production of recombinant DNA. The third stage of a genetic engineering experiment consists of a plasmid with DNA segments that can be placed into the other bacterial cell. Now when this cell reproduces, it produces an exact clone of the DNA segments introduced by the plasmid, which was just mentioned. The plasmid with DNA segments is referred to as the vector. Screening is stage four of the genetic cloning experiment. This consists of looking for a specific DNA segment that is associated with the original vector that was introduced.

71. A _____ is referred to as the genome that brings the unrelated DNA into the host cell in genetic engineering. Thus, _____ is a kind of DNA that is made in the laboratory.

*   a. vector, recombinant DNA
        b. vector, DNA polymerase
        c. transposon, recombinant DNA
        d. transposon, DNA polymerase

72. Hybridization is used in the _____ stage of genetic engineering experiments.
        a. DNA cleavage
        b. production of recombinant DNA
        c. cloning
*   d. screeening
Hybridization is used to screen for a particular gene. This procedure involves cloned genes that connect by producing base pairs with another complementary sequence on another piece of DNA.

73. What is used to specifically cleave DNA in genetic engineering experiments?
        a. transposon
        b. plasmid
*   c. restriction endonuclease
        d. hybridization

74. What is the complementary nucleic acid referred to in the hybridization used in the screening stage of a genetic engineering experiment?
        a. restriction endonuclease
*   b. probe
        c. plasmid
        d. transposon
The probe is used to find out the presence of a certain gene in the clone.

75. _____ is the first stage in a genetic engineering experiment. This stage uses restriction endonuclease to cut or cleave the DNA into segments.
        a. cloning
*   b. DNA cleavage
        c. production of recombinant DNA
        d. screening

76. Which of the following is a step in the polymerase chain reaction?

a. denaturation

b. annealing of primers

c. primer extension

*   d. all of the above

In step one, denaturation of DNA occurs. A small DNA segment will form into single strands. The denaturation occurs by the use of heat and a primer. The primer is mixed with the small DNA segment with heat. In step two, the solution is cooled down and the primers connect with the complementary DNA single strands. Step two is called the annealing of primers. In step three, DNA polymerase copies each strand. This makes two replicated strands. These steps are repeated many times. The whole purpose of the polymerase chain reaction (PCR) is to enlarge small segments of DNA into larger segments that later can be studied.

77. The sections of DNA that were cleaved by restriction endonuclease in the DNA cleavage stage of a genetic engineering experiment will then be placed into a plasmid in stage _____. This is called the stage of _____.

*   a. two, production of recombinant DNA

b. four, screening

c. three, cloning

d. one, DNA cleavage

78. Which of the following techniques is used to identify a specific person at a scene of an investigation?

a. PCR

*   b. RFLP analysis

c. Southern blot

d. all of the above

Restriction fragment length polymorphism (RFLP) analysis is used to identify a specific person using a certain gene as an identification determination. This is done by cleaving a piece of a person's DNA with restriction endonuclease. The cut or cleaved segments will result. However, remember that every individual will have different measurements in the lengths of these segments or fragments of DNA. Gel electrophoresis is then used to separate these fragments of DNA. In addition, varying banding patterns will be seen and are different in each individual.

79. Which of the following is a type of gene technology?
   a. Southern blot
   b. PCR
   c. RFLP analysis
*   d. all of the above

80.   In which method of gene technology is DNA cleaved into smaller pieces by restriction endonuclease and the cut segments are pulled apart by electrophoresis?
   a. hybridization
   b. Sanger method
   c. PCR
*   d. Southern blot
The Sanger method is used to spot and identify DNA sequences that are not known. Remember that the Southern blot and the Sanger method both use gel electrophoresis.

81.   Denaturation, primer extension and annealing of primers are the three procedures in which of the following types of gene technology?
   a. Southern Blotting
*   b. polymerase chain reaction (PCR)
   c. Sanger method
   d. restriction fragment length polymorphism (RFLP)